CELLS

THE BUILDING BLOCKS OF LIFE

How Scientists
Research Cells

Cells: The Building Blocks of Life

CELLS
THE BUILDING
BLOCKS OF LIFE

How Scientists
Research Cells

KRISTI LEW

CHELSEA HOUSE
An Infobase Learning Company

Chelsea House
An imprint of Infobase Learning
132 West 31st Street
New York NY 10001

Library of Congress Cataloging-in-Publication Data
Lew, Kristi.
  How scientists research cells / by Kristi Lew.
      p. cm. — (Cells, the building blocks of life)
  Includes bibliographical references and index.
  ISBN 978-1-61753-007-4 (hardcover)
  1. Cells—Popular works. I. Title. II. Series.
  QH582.14.L45 2011
  571.6—dc22   2011004475

Chelsea House books are available at special discounts when purchased in bulk quantities for businesses, associations, institutions, or sales promotions. Please call our Special Sales Department in New York at (212) 967-8800 or (800) 322-8755.

You can find Chelsea House on the World Wide Web at http://www.infobaselearning.com

Text design by Erika K. Arroyo
Cover design by Alicia Post
Composition by EJB Publishing Services
Cover printed by Yurchak Printing, Landisville, Pa.
Book printed and bound by Yurchak Printing, Landisville, Pa.
Date printed: August 2011
Printed in the United States of America

10 9 8 7 6 5 4 3 2 1

This book is printed on acid-free paper.

All links and Web addresses were checked and verified to be correct at the time of publication. Because of the dynamic nature of the Web, some addresses and links may have changed since publication and may no longer be valid.

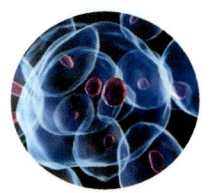

Contents

● ● ●

The Discovery of Cells

All living things are made up of cells. Some living organisms, such as bacteria and protozoa, are composed of only one cell. Other organisms, such as human beings, have trillions of cells. Cells are the building blocks of life. Therefore, scientists seeking to understand how living systems work often start by studying the composition, function, and interaction of these tiny, complex structures.

The road to discovering the cell was not an easy one. Cells are too small to be seen without the use of a microscope. Before the microscope's invention, scientists were not aware that cells existed. The technology that would eventually reveal these tiny building blocks was invented in the second half of the seventeenth century.

EARLY OBSERVATIONS

Englishman Robert Hooke (1635–1703) was one of the best experimental scientists of his time. Hooke was a multitalented inventor, microscopist, physicist, biologist, astronomer, surveyor, and artist, and he made contributions to many scientific disciplines. Among other things, Hooke invented the universal joint (without which automobile drive shaft systems would not work), a mechanism (called an anchor escapement) that allowed more accurate clocks to be built, and a better microscope.

Hooke survived smallpox as a child, but he was permanently scarred by the disease. Too sick to go to school, he was mainly taught by his

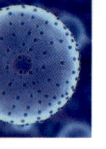

FIGURE 1.1 British scientist Robert Hooke made numerous scientific discoveries, yet there are no portraits of him. Thus, this bust—from the Hooke Museum on the Isle of Wight in the United Kingdom—was based only on written descriptions of him. While there is some controversy over why this is the case, some historians suggest that fellow British scientist Isaac Newton, who died 24 years later, attempted to eradicate Hooke's likeness from history due to their long rivalry over studies in light and gravitation, among other things.

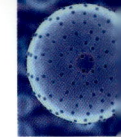

father. He was also a naturally curious child and interested in a variety of scientific pursuits. He lived on the Isle of Wight, an island off of the coast of England. The Isle of Wight has many fossils and different habitats. This sparked Hooke's interest in paleontology and biology. After his father's death, 13-year-old Hooke left the Isle of Wight and traveled to London to attend Westminster School and briefly serve as an apprentice to an artist named Peter Lely. In 1663, Hooke graduated with a master's degree from Oxford University at the age of 26.

Hooke did not invent the microscope, but he did improve on an existing design. Using his newly improved microscope, Hooke observed insects, sponges, bryozoans, foraminifera, and bird feathers. Using his training as an artist, he drew detailed illustrations of the objects he observed. These detailed drawings and written observations were compiled in a book called *Micrographia,* which was published in 1665. People were fascinated with Hooke's observations and *Micrographia* quickly became a bestseller.

One of Hooke's more enduring findings—and the one that he is probably most remembered for—is his observations of a thin slice of cork. Looking through his newly designed microscope, Hooke saw that the cork was made up of a network of spaces that looked like an irregular honeycomb. The small spaces reminded Hooke of the small rooms the monks at the monastery lived in. The Latin word *cella* means "small room," so Hooke called the small, square compartments in the piece of cork cells.

While Hooke improved on the existing microscope design and was the first to describe plant cells, one of his contemporaries, Antonie van Leeuwenhoek (1632–1723), was making his own observations and improving microscope lenses. Unlike Hooke, van Leeuwenhoek did not set out to be a scientist. In fact, he was a clothing and dry goods salesman and had no formal scientific training. However, he and Hooke did have something in common—they both shared a seemingly endless curiosity about the world around them. Van Leeuwenhoek became fascinated with the magnifying glasses of the day and started making some of his own lenses. In time, he became so adept at making and polishing clear glass magnifying lenses that he was able to produce a lens that could enlarge objects more than 200 times their natural size. Using his lenses, van Leeuwenhoek observed blood cells and spermatozoa. He also observed and described single-celled organisms he called *animalcules.* Scientists today call these creatures *bacteria* and *protists.* Van Leeuwenhoek's observations of bacteria and other microorganisms were some of the first ever recorded.

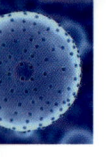

THE MICROSCOPE IMPROVES

During the eighteenth and nineteenth centuries, microscope technology continued to improve. One key improvement was made by Joseph Jackson

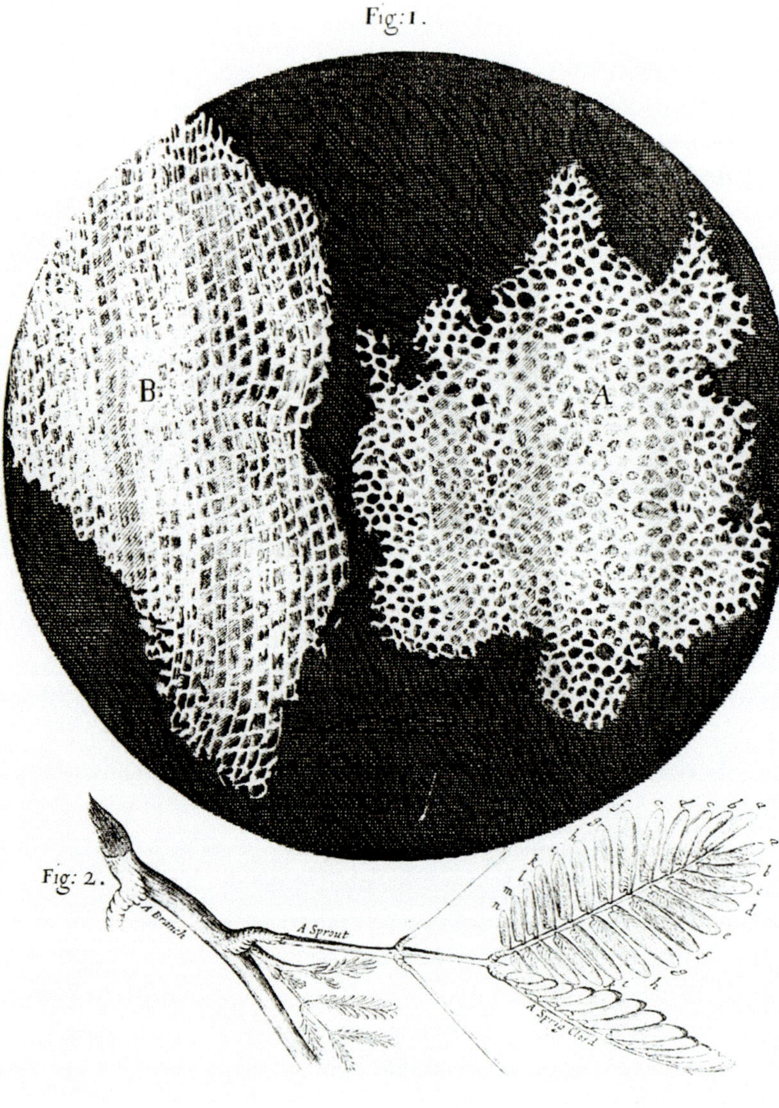

Fig: 1.

B.

A.

Fig: 2.

A Branch

A Sprout

A Sprig Cork

FIGURE 1.2 Robert Hooke's sketch observations of the cellular structure of cork and a sprig of "sensible sensitive plant" were created in 1665 after he viewed the objects under a compound microscope. They were published in his book *Micrographia*. Hooke was the first to use the word *cell* to describe the honeycomb nature of cork.

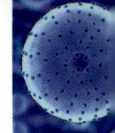

Lister (1786–1869), a British optics specialist. Lister started experimenting with microscope lenses in the mid-1820s. In 1830, he announced that he had discovered a way to combine microscope lenses that eliminated much of the distortion that was present before. Lister's microscope, called an achromatic microscope, contained two weak lenses set at a particular distance from one another. Each lens was made from a different type of glass. The different types of glass helped to counteract the distortion caused by the bending of light rays as they passed through the lenses. Without the distortion, scientists could see cells (and other microscopic objects) more clearly than ever before. Lister's lens-making technique soon became an industry standard for microscope makers. Over time, more advances in technology allowed microscopes to become smaller and, therefore, easier to handle, thereby making them more popular with scientists.

Now that scientists were able to see more clearly, formerly undetectable details were quickly noted. In 1831, Robert Brown (1773–1858), a Scottish botanist, first described the cell **nucleus** in a paper he presented to the Linnean Society (a leading biological sciences organization). He found that all of the cells in the orchid leaves he was studying contained a small dark spot. Although Brown probably was not the first scientist to see these dark spots, he was the first to describe them in a scientific paper. He also recognized the nucleus as a necessary component of a cell. Brown's paper was published in 1833. Though Brown is often given the credit for finding the cell nucleus, he always credited another botanist, Franz Bauer (1758–1840), for making observations similar to his own at the same time.

CELL THEORY

Five years after Brown's observations were published, Matthias Jakob Schleiden (1804–1881), a German botanist, proposed the idea that all plants are made up of cells. A year later, in 1839, German zoologist Theodor Schwann (1810–1882) suggested the same idea about animals.

In 1852, Robert Remak (1815–1865), a Polish-German embryologist (a scientist who studies the development of a fertilized egg into a fetus), developed a method of staining cells that allowed him to see how cells divide. As a result of his observations, Remak concluded that cells originate from the division of other cells. He also theorized that disease was the result of changes in an organism's cells.

Six years later, a German scientist named Rudolf Virchow (1821–1902) read Remak's scientific papers and set out to popularize his ideas. Virchow also came to the conclusion that cells originate from other cells and that disease is the result of changes in an organism's cells. Virchow expanded and explained Remak's ideas in a book called *Cellularpathologie* (*Cellular Pathology*), which was published in 1858. Today, the idea that all living things are made up of cells and that all cells originate from other cells is known as the **cell theory**.

By the middle of the nineteenth century, scientists knew that cells were made up of a cell wall or membrane, a fluid called protoplasm (the name was changed to cytoplasm in 1874, the term given to it by Rudolf Albert von Kolliker [1817–1905]), and a nucleus.

HOW SCIENTISTS WORK: THE SCIENTIFIC METHOD

People often view the world differently. However, scientists cannot allow their observations and conclusions to be influenced by their personal and cultural beliefs. To minimize these personal biases, scientists use a set of techniques collectively known as the scientific method. Scientists use the scientific method in an attempt to develop reliable, consistent theories that accurately describe natural phenomena. The scientific method has several stages:

- **Collect observations.** Scientists are usually good observers. They notice what is going on in the world around them and become curious about these observations. A scientist may make observations directly, by reading and studying what other scientists have done in the past, or by talking to their colleagues.
- **Define the question.** Observations often lead to questions. Scientific questions are testable in some way. Depending on what observation is being studied, stages one and two may be switched.
- **Form a hypothesis.** A hypothesis is testable, possible answer to the question.
- **Test hypothesis by collecting data.** There are several ways to collect scientific data. One way is through experimentation. Scientific experiments are repeatable. However, not all scientists perform experiments. Paleontologists, for example, cannot experiment on dinosaurs to prove their hypotheses. Instead, they gather data by

DISCOVERING ORGANELLES

The discovery of cell **organelles** took a bit longer. Organelles are specialized parts inside the cell where particular chemical reactions necessary to cell function are carried out. **Chloroplasts**, for example, are plant organelles that turn light energy into chemical energy during a process called photosynthesis. The chemical energy created during photosynthesis is stored in the form of sugar. **Mitochondria**, which are cell organelles that are found in both plants and animals, act as the cell's digestive system. Mitochondria break down sugar and other nutrients, converting them into the energy that cells need in order to function correctly.

observing dinosaur fossils. Other scientists use modeling to test their hypotheses. Weather forecasts are an everyday example of scientific modeling.

- **Organize and analyze data.** Scientists often build on their own work or on the work of others. Therefore, data collection should occur systematically and be recorded in such a way that other scientists can understand how it was gathered. This type of data collection allows other scientists to repeat experiments and validate the results.
- **Communicate conclusions.** Sharing the results of experimentation or observation is an important step in the scientific method. Many scientists publish their results in scientific journals. This allows other scientists to read, study, and build upon their work.
- **Retest hypothesis.** This step is often done by other scientists.

After repeated testing, a hypothesis may become a theory. In everyday speech, a theory can mean an opinion or a guess. However, in science, this is not true. A scientific theory is an explanation that has been tested over and over again through repeated experimentation and observation. Scientists generally accept theories as fact until they are proven otherwise by well-constructed experiments and validated data.

Science does not always move in linear fashion and the stages of the scientific method are not always done in the exact order listed above. However, a properly executed scientific inquiry contains most, if not all, of these stages.

The discovery of mitochondria is not attributed to a single scientist, but to several. These scientists worked and studied mitochondria for over a century and a half in order to refine its structure and function. In 1857, Rudolf Albert von Kolliker, a Swiss scientist, was the first to report the presence of small specks, or granules, in muscle cells. Other scientists noticed these granules in other cells, too. In 1886, Richard Altmann

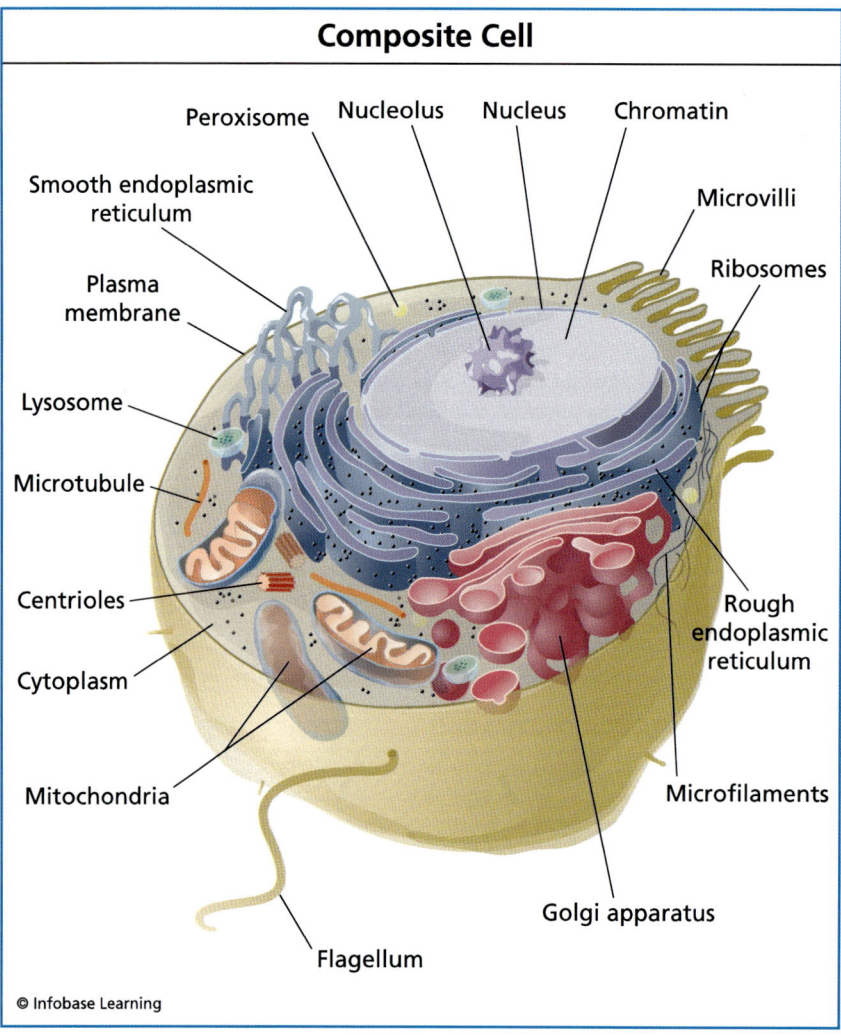

Composite Cell

Peroxisome Nucleolus Nucleus Chromatin

Smooth endoplasmic reticulum

Microvilli

Plasma membrane

Ribosomes

Lysosome

Microtubule

Centrioles

Rough endoplasmic reticulum

Cytoplasm

Mitochondria

Microfilaments

Golgi apparatus

Flagellum

© Infobase Learning

FIGURE 1.3 This composite cell illustrates some of the common features and organization of many cell types. However, it does not give an idea of the great diversity in size, shape, and structure among cells, which reflects their different functions.

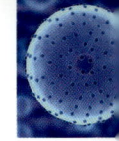

(1852–1900), a German pathologist, developed a way to stain cells that allowed him to observe these granules more closely. Altmann called the structures "bioblasts." In 1898, Carl Benda (1857–1933) renamed them mitochondria. Over the next century, many other scientists would be involved in describing the structure and function of these very important cellular components.

Mitochondria are not the only cell organelles that were being investigated at that time. Scientists were also studying the structure and the content of the cell nucleus. In the late 1800s, Johannes Friedrich Miescher (1844–1895), a Swiss physician and biologist, became interested in the chemistry of bodily fluids, particularly that of blood. To further his research, Miescher went in search of easily obtained cells. He successfully found a large quantity of white blood cells, one of the main components of pus, at a nearby hospital where wounded soldiers from the Crimean War were being treated. In 1869, Miescher successfully isolated the nuclei from the white blood cells and, from these nuclei, a new substance he called nuclein. In 1874, Miescher separated nuclein into a protein and an acid molecule. In 1889, one of Miescher's students, Richard Altmann (the same Richard Altmann who developed the staining technique that revealed the existence of mitochondria in 1886) renamed the molecule nucleic acid. Building on Miescher's work, other scientists would later discover more about the structure of this molecule and its name was changed again. Today, it is known as **deoxyribonucleic acid**, or **DNA**. As early as 1869, Miescher proposed that the threadlike molecule might have something to do with heredity, but it would take scientists 75 years to confirm this idea.

One scientist who contributed to the elucidation of DNA's function in the body was a German biologist named Walther Flemming (1843–1905). In 1879, Flemming developed a new staining method that allowed him to see structures in the nucleus that he called chromatin. His new dye stained these structures a deep red. The word *chromatin* is derived from a Greek word that means "color." He also noticed that the threadlike material condensed into structures that moved and separated during cell division, a process that Flemming named mitosis. Today scientists call these condensed structures chromosomes.

Flemming was not the first scientist to describe cell division. This distinction belongs to the Swiss botanist Karl Wilhelm von Nägeli (1817–1891). Nageli observed the process in pollen grains as early as 1842. However, he thought the process was an anomaly. Nägeli's observations of cell division in plants and Flemming's observations of the same process in cells of salamander embryos helped scientists to better understand the connection between all living things.

While research into the structure and function of the mitochondria and nucleus was ongoing, other scientists turned their attention to other

ENDOSYMBIOTIC THEORY

Prokaryotes are single-celled organisms that do not have a nucleus or other organelles. Bacteria are prokaryotes. The endosymbiotic theory describes one way scientists believe that organelles may have evolved. This theory proposes that some eukaryotic organelles, such as the mitochondria and chloroplasts, were at one time separate organisms. In other words, it is possible that in the distant past, mitochondria and chloroplasts were some type of primitive bacteria.

The fact that these organelles have their own DNA is one compelling reason for scientists to believe that mitochondria and chloroplasts may have started out as separate organisms. In addition, the DNA found in mitochondria and chloroplasts is circular. Prokaryotic DNA is circular, too. The organelles also have their own ribosomes. The ribosomes in mitochondria and chloroplasts are slightly smaller than the cellular ribosomes found in the cytoplasm. The same is true of prokaryotic ribosomes, which are slightly smaller than the ribosomes found in the cytoplasm of eukaryotes.

In 1905, Konstantin Mereschkowsky (1855–1921), a Russian botanist, became the first to suggest that chloroplasts began as separate organisms. Lynn Margulis (1938–), an American biologist, studied and expanded on Mereschkowsky's work and published her hypothesis in *The Origin of Eukaryotic Cells* in 1970. Today, most biologists accept Margulis's theory that says that a few billion years ago, a primitive eukaryote engulfed, but did not digest, a primitive prokaryote. Over time, the eukaryote and the prokaryote evolved together in a symbiotic relationship (a relationship that benefits both) until neither one could live independently of the other.

Opposite: Figure 1.4 The endosymbiotic theory purports that mitochondria and chloroplasts in eukaryotic cells originated from smaller prokaryotic cells living inside larger cells.

cell organelles. For example, Camillo Golgi (1843–1926), an Italian scientist, was a pioneer in the discovery of how the nervous system worked. To effectively study neurons (nerve cells), Golgi developed a staining technique that still bears his name today. Golgi staining (sometimes called

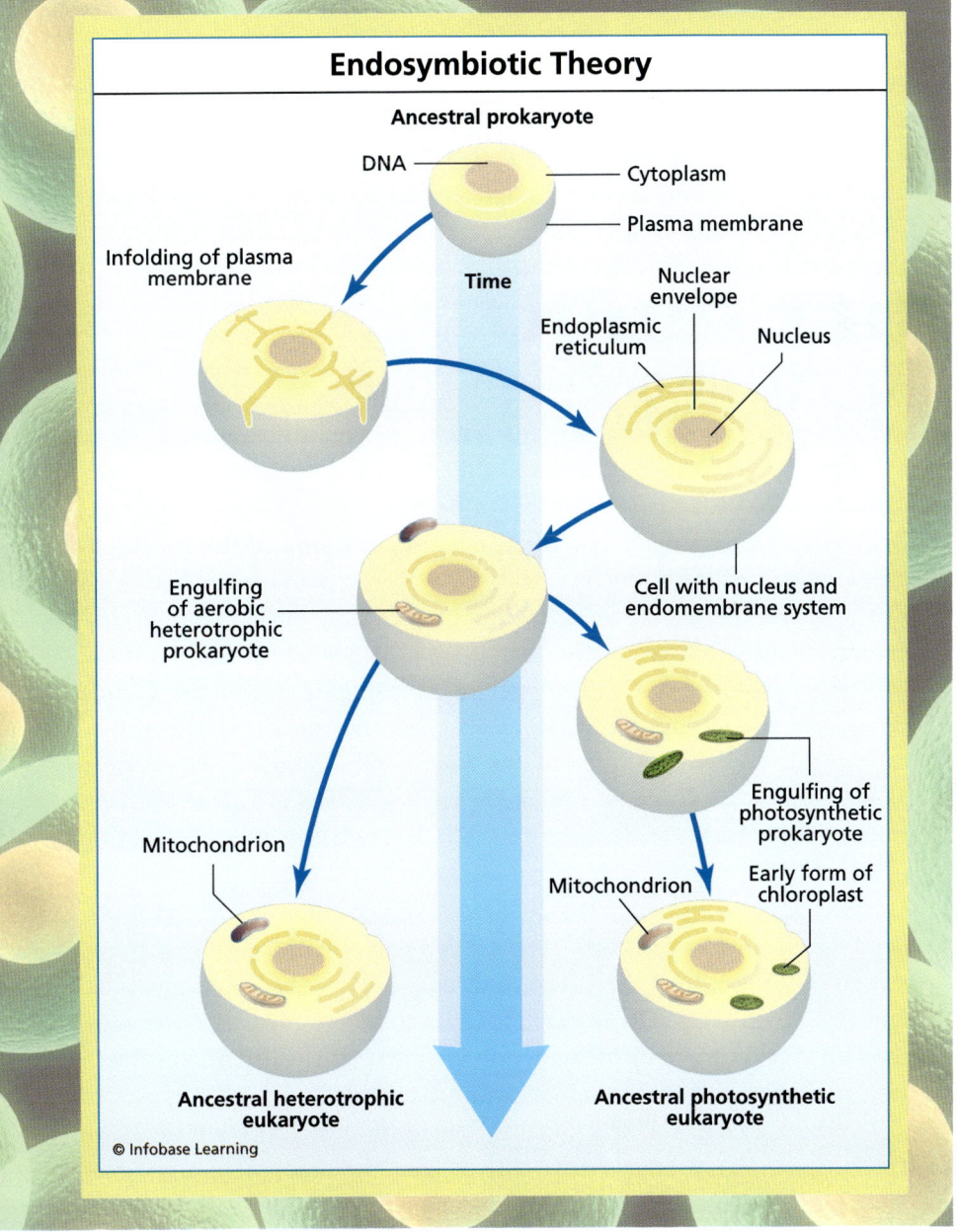

Endosymbiotic Theory

Ancestral prokaryote

DNA

Cytoplasm

Plasma membrane

Infolding of plasma membrane

Time

Nuclear envelope

Endoplasmic reticulum

Nucleus

Engulfing of aerobic heterotrophic prokaryote

Cell with nucleus and endomembrane system

Engulfing of photosynthetic prokaryote

Mitochondrion

Mitochondrion

Early form of chloroplast

Ancestral heterotrophic eukaryote

Ancestral photosynthetic eukaryote

© Infobase Learning

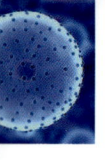

Golgi impregnation) revealed an internal cell structure that Golgi named the "internal reticular apparatus." The name was eventually changed to the Golgi apparatus and, today, is known as the **Golgi complex**, the Golgi body, or simply "the Golgi." The Golgi complex stores, sorts, and packages proteins. These proteins may be used within the cell for various functions or they may be excreted by the cell. Golgi bodies are found in both plant and animal cells.

Not all scientists believed that what Golgi was seeing was actually a cell organelle. Many believed that it was a staining artifact. An artifact is something that looks like a structural detail, but is, in fact, an anomaly caused by the staining process. Whether the Golgi apparatus was in fact a newly discovered organelle or simply an artifact of staining was debated in the scientific community for decades. The existence of the Golgi apparatus was finally confirmed in the mid-1950s with the use of an electron microscope.

NEW TECHNIQUES AND TOOLS

To discover more cell organelles, scientists needed different techniques and tools. One of these techniques, called cell **fractionation**, was developed by Albert Claude (1899–1983), a Belgian-born biologist working at Rockefeller University in New York. Cell fractionation makes use of high-speed centrifugation, or ultracentrifugation. When cells are spun at a high rate of speed, the cell breaks apart and heavier components become concentrated in the bottom of the centrifuge tube. Lighter particles, on the other hand, come to rest in the upper layers. Using this technique, Claude discovered the **endoplasmic reticulum**, the interconnected network of tubules that serves as an internal transport system in the cell. In the early 1940s, Claude also became one of the first scientists to use an electron microscope, which was developed in 1931, to study the interior structure of cells.

George Emil Palade (1912–2008), one of Claude's students, also utilized the electron microscope. He used this new scientific tool to map the structure of the mitochondria and study chloroplasts, the Golgi apparatus, and other organelles. His research led him to determine that ribosomes, thought at that time to be pieces of mitochondria (and called microsomes), were actually a part of the endoplasmic reticulum.

Using Claude's cell fractionalization technique, Christian de Duve (1917–), a Belgian biochemist, noticed that the amount of a particular enzyme released by the cell varied in direct relation to how much

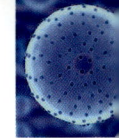

damage was done to the cell while it was being centrifuged. De Duve hypothesized that this enzyme must be contained within the membrane of an organelle. He calculated the approximate size of these kinds of organelles and named them lysosomes. Lysosomes function as the digestive organelle of the cell. De Duve later identified lysosomes within the cell using the electron microscope. His discovery explained how the cell separated the digestive enzymes it needs to break down nutrients from other parts of the cell that should not be digested. Claude, Palade, and de Duve were awarded the 1974 Nobel Prize for Physiology or Medicine for their work.

The use of the electron microscope to study individual organelles, as well as cells as a whole, eventually allowed scientists to discover how each organelle's shape and placement in the cell allowed it to carry out its biological function.

In the early 1960s, cell fractionalization revealed that the nucleus might not be the only organelle that contained DNA. In 1962, Hans Ris (1914–2004) and Walter Plaut used specialized staining techniques to confirm that chloroplasts contained their own DNA. The staining procedure used by Ris and Plaut was designed for bacterial cells. The fact that the staining was originally intended for bacteria and that chloroplasts had DNA as bacteria do, gave Lynn Margulis, who at the time was a graduate student in Plaut's laboratory, her idea about the endosymbiotic theory. Ris and Plaut later used an electron microscope to visualize chloroplast DNA, confirming their staining results.

A year after Ris and Plaut discovered chloroplast DNA, the husband and wife research team of Margit Nass-Edelson and Sylvan Nass utilized the electron microscope to show that mitochondria also contain DNA. Their discovery was confirmed by biochemical tests a year later by Ellen Haslbrunner, Hans Tuppy, and Gottfried Schatz.

A variety of scientists are still studying the cell today. The same tool that allowed Hooke to visualize the very first cell—the light microscope—is still used in biology laboratories around the world. In addition, the electron microscope and various staining techniques have allowed scientists over the years to see ever smaller details to add to the overall picture of a living cell. New techniques designed to keep cells alive outside of the body and the ability to track molecules within the cell have also given scientists the tools they need to ascertain how different parts of the cell function and how cells work together to keep more complex organisms, such as humans, alive and well.

2

Microscopy

The microscope is one of the main tools used by cell researchers. The light, or optical, microscope uses light to illuminate the object being seen. The electron microscope, invented in the 1930s, uses electrons to render an image of objects that are too small to be seen with a light microscope.

HISTORY OF THE MICROSCOPE

Around the year A.D. 100, Romans discovered a way to make glass transparent by adding the chemical manganese oxide to the glassmaking process. While experimenting with this new material, they discovered that if they made the glass thick in the middle and thinner around the edges, they could use it to direct and focus the Sun's rays. They also discovered that focusing the Sun's rays in this manner could set objects on fire. Appropriately, they used the term *burning glasses* to describe glass pieces with this convex shape.

Burning glasses also had another interesting function. When an object was viewed through one of them, it appeared larger than it did to the naked eye. These glass pieces were, essentially, the first magnifying glasses.

Later, in about the thirteenth century, philosopher and renowned scientist Roger Bacon (1214–1294) concluded that convex glass lenses could be used to aid people who had trouble seeing. The first spectacles, or

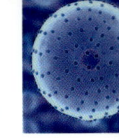

FIGURE 2.1 Dutch spectaclemaker Zacharias Janssen (*seen here*) and his father are credited with inventing the first optical telescope, and, arguably, the first true compound microscope.

eyeglasses, for correcting farsighted vision were reported in Italy and China in the late 1200s. Glasses to correct nearsightedness, a condition called myopia, have lenses that are shaped differently; they did not appear until the 1400s.

In the late sixteenth century, Dutch eyeglass maker Zacharias Janssen (1580–1638) and his father, Hans, discovered that looking at an object through more than one glass lens could greatly increase the magnification of the object. The apparatus that the Janssens invented was comprised of two lenses mounted at opposite ends of a tube. This piece of equipment is believed to be one of the first compound microscopes ever made. An important tool still used by today's scientists, the compound microscope is a microscope that is made up of two or more lenses.

Earlier in the sixteenth century, another Dutch optician, Hans Lippershey (1570–1619), also put two glass lenses into either end of a long tube in order to invent the spyglass, a precursor of the telescope. One of the main differences between a telescope and a microscope is that the lens in the eyepiece of a telescope is smaller than the lens on the opposite end of the tube. The primary function of this large lens is to gather light from distant objects. From the descriptions of the first microscopes, some historians believe that they may have been based on looking through the wrong end of a telescope. When the image of a distant object passes through the larger lens and then the smaller eyepiece, it is greatly magnified.

One of the more well-known scientists who experimented widely with the earliest telescope was Italian physicist and astronomer Galileo Galilei (1564–1642). Galileo not only used the telescope to strengthen Italian astronomer Nicolaus Copernicus's idea that Earth and other planets revolved around the Sun, he also conducted further research on lenses and light rays. Using his observations on how glass lenses and light rays functioned, he went on to make many improvements to both the telescope and the microscope which included the first focusing devices.

Early microscopes were very simple. They were not much more than basic magnifying glasses that could enlarge objects six to ten times their natural size. Nevertheless, scientists and others used these early tools to observe objects, such as fleas and other tiny insects, that were too small to be seen without them. One scientist who was especially interested in microscopy was Antonie van Leeuwenhoek of the Netherlands.

Van Leeuwenhoek did not start out as a scientist. Although he was fluent in his native Dutch, he had no higher education and did not know English or Latin (the language scientific papers were published in at the time). However, he did possess a great curiosity. While working in a dry goods store, he became fascinated with magnifying glasses. He used these lenses to observe many things, including the thread count of woven cloth.

Soon, he learned to grind and polish his own lenses and discovered that lenses with the greatest curvature produced the most magnification. Using this knowledge, he was able to produce a lens that could magnify an object 270 times its natural size. With this magnification at his disposal,

FIGURE 2.2 While using handcrafted microscopes, Antonie van Leeuwenhoek became the first person to observe and describe single-celled organisms, which are now referred to as microorganisms. For this reason, he is commonly referred to as the father of microbiology.

van Leeuwenhoek went on to discover and describe bacteria, yeast, spermatozoa, blood cells, protozoa. and nematodes (microscopic worms) for the first time.

Hooke's Microscope

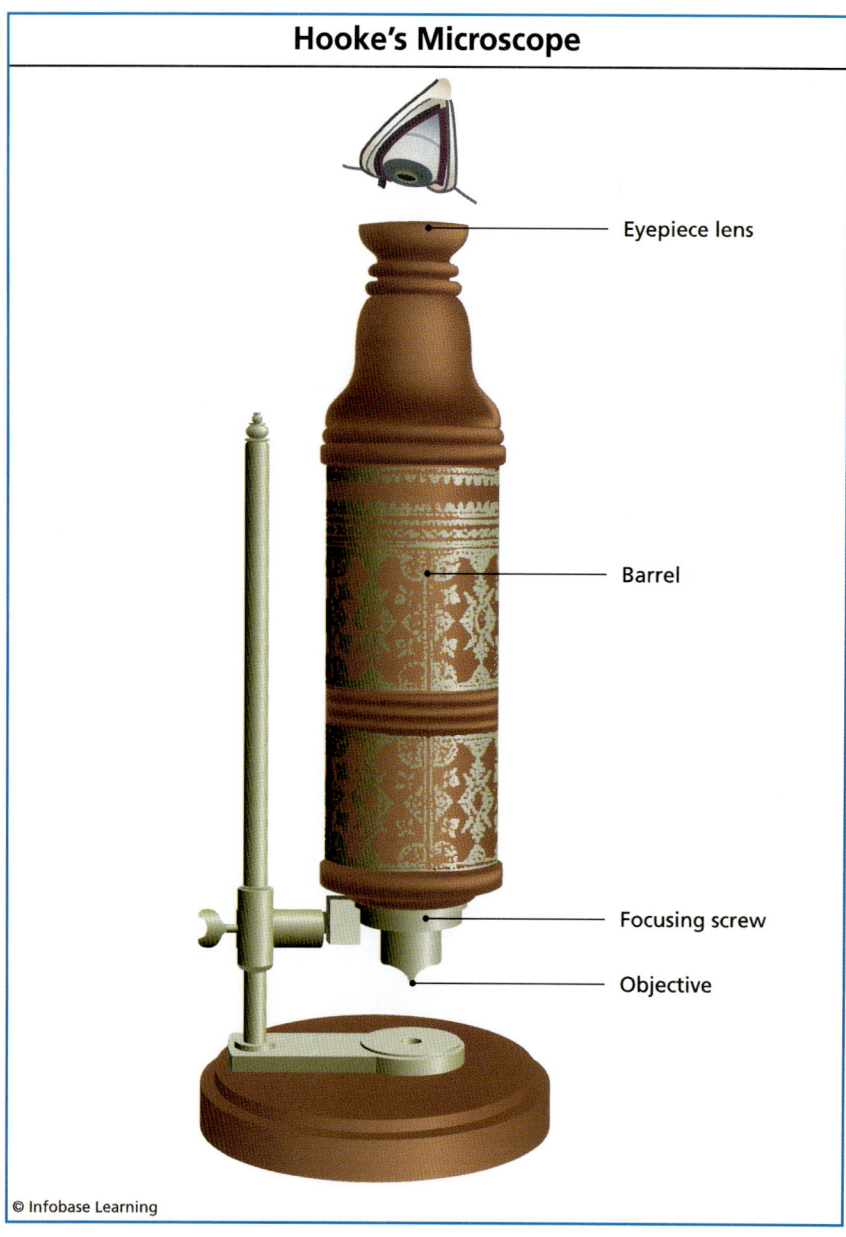

Eyepiece lens

Barrel

Focusing screw

Objective

© Infobase Learning

FIGURE 2.3 The microscope Robert Hooke designed circa 1665 was 6 inches (15 centimeters) long. Although its craftsmanship was said to be excellent, its fragile focusing mechanism would wear out fairly quickly.

Some historians believe that van Leeuwenhoek may have been inspired to carry out his observations by another scientist, Robert Hooke. Hooke's illustrated *Micrographia,* which detailed his microscopic observations of many objects including fleas, spiders, and mites, was published in 1665. This book was extremely popular. Yet Hooke was not only a writer and illustrator. At the tender age of 13, he quickly distinguished himself at Oxford University as a scientist with an excellent sense of experimental design. He was also talented in the construction and improvement of scientific equipment. One piece of scientific equipment that Hooke studied, experimented with, and redesigned was the compound microscope. He also designed a lighting system that would allow specimens to be seen more clearly.

With the improved microscope, Hooke observed insects, sponges, foraminifera (single-celled protists with shells), bird feathers, and many other objects, including a thin slice of cork. He noticed that the cork was made up mostly of air, which explained why it would float on water. He also observed that a meshlike structure enclosed the pockets of air. To him, the pockets of air looked like the tiny rooms in a monastery, called cells. Therefore, Hooke also named these pockets of air *cells.* Although he did not know it at the time, he was observing plant cells and cell walls for the first time.

During the eighteenth and nineteenth centuries, scientists continued to improve on the microscope. Microscope slides were developed and flat platforms, called stages, were added to microscopes to improve the stability of samples. Improvements in focusing, magnification, and the development of staining techniques also improved the ability of scientists to observe objects they could not see without these new and improved scientific tools.

THE COMPOUND LIGHT MICROSCOPE

Compound microscopes are sometimes called optical microscopes. Most common optical microscopes use light to illuminate the object being studied. Therefore, these microscopes are also called light microscopes.

As discussed earlier, a compound microscope has two or more glass lenses. One lens is in the eyepiece of the microscope and is called the **ocular lens**. This lens usually has the power to magnify an object 10 (10x) to 15 times (15x) its natural size. The other lens, called the **objective lens**, is closer to the sample that is being observed. The two lenses are fixed on either end of a closed tube. Together, the lenses can achieve much more magnification than one lens alone. Many modern optical microscopes

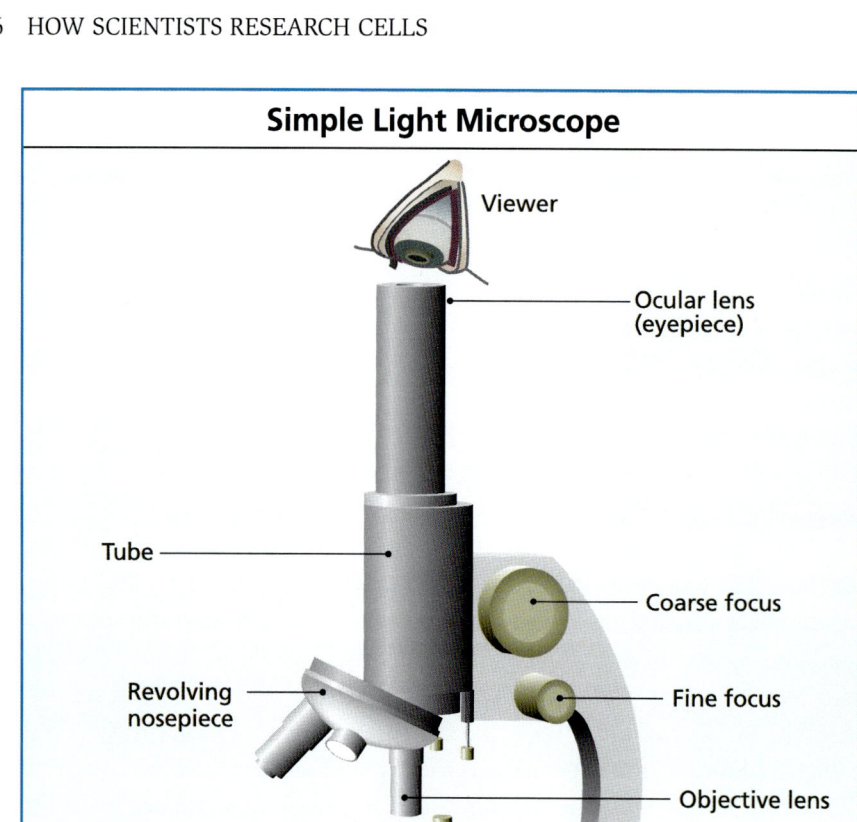

Simple Light Microscope

Viewer

Ocular lens (eyepiece)

Tube

Coarse focus

Revolving nosepiece

Fine focus

Objective lens

Stage with clips

Arm

Light source

Base

© Infobase Learning

FIGURE 2.4 The simplest modern light microscope is a sophisticated instrument. The eyepiece lens (usual magnification 10x or 15x) is mounted at one end of the tube. At the other end, the revolving turret carries two or more objective lenses (usual magnifications 4x, 10x, 40x, and 100x). The total magnification is that of the objective lens multiplied by the eyepiece lens. The arm supports the tube. The slide fits on the stage; clips prevent it from moving about. A light source shines upward through the slide; some microscopes use a mirror to reflect ordinary light.

have three or more objective lenses of varying degrees of magnification. These objective lenses are often called low-, medium-, and high-powered lenses. These lenses are attached to a circular wheel called the nosepiece. The nosepiece is turned to allow the observer to change from low to medium or high power.

The magnification power of a compound microscope is determined by multiplying the power of the ocular lens and the power of the objective lens. If the eyepiece magnifies an object by 10x power and the objective lens by 40x power, for example, the total magnification of the specimen is 400 times its natural size.

The sample to be viewed is placed on a flat plate below the objective lenses called the stage. To bring the object being studied into focus, it is moved closer to or farther away from the objective lens. This movement can either be accomplished by moving the tube that contains the lenses or by moving the stage up and down. This movement is controlled by focusing control knobs on the side of the microscope.

Below the stage is the light source. Some light microscopes have an electric lamp that shines light through a hole in the stage and illuminates the specimen. Other microscopes have a mirror. The mirror can be adjusted to reflect daylight or lamplight through the hole in the stage. More sophisticated microscopes may also have a round apparatus called the condenser below the stage. The condenser's job is to focus the light onto the specimen. Some light microscopes also have a rotating disk, called a diaphragm, under the stage. The diaphragm controls the amount of light that comes through the condenser. Settings on the diaphragm and condenser will depend upon the transparency of the specimen and how much contrast is desired for viewing.

PREPARING AND VIEWING SAMPLES

Specimens need to be prepared before they can be viewed under a microscope. Preparation methods include dry mounts, wet mounts, and prepared mounts. All of these preparation methods result in the same thing—they position the sample on a microscope slide. A microscope slide is a small, rectangular piece of glass on which the sample is placed. A coverslip, which is a very thin circular, square, or rectangular piece of transparent glass or plastic, is placed on top. The coverslip prevents the sample from touching and harming the objective lens. It also protects the specimen from being damaged.

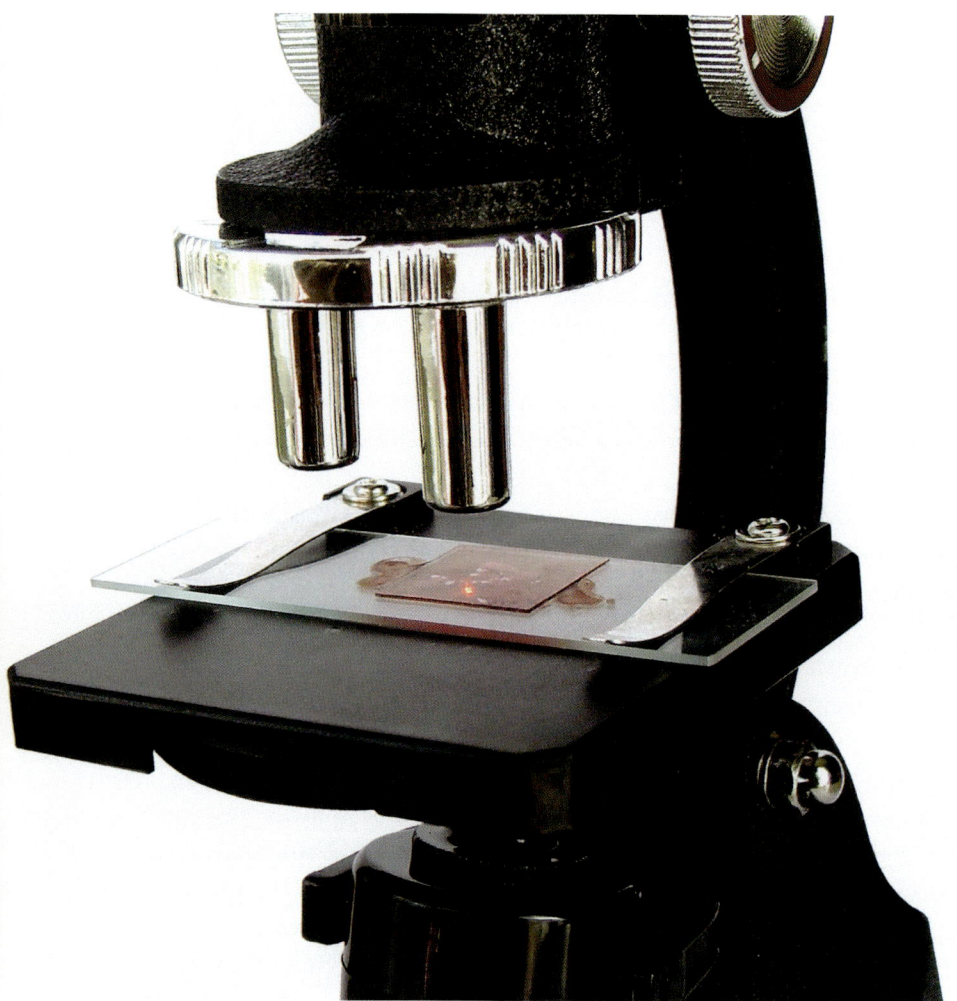

FIGURE 2.5 A coverslip, sometimes called a cover glass, is a flat, transparent circle, square, or rectangle that is placed over the viewing objects on a microscope. It typically rests on a slightly thicker glass slide, with the microscope stage supporting from underneath.

A dry mount does not require the use of water or other liquid to prepare. The object to be viewed is placed on top of a microscope slide and covered with a coverslip. However, it is important that the specimen be very thin; if possible, only one cell layer thick. If the specimen is too thick, light will not be able to penetrate it, making it hard to see. Placing a layer of onion skin or a single human hair on a microscope slide and covering it with a coverslip are examples of dry mounts.

A wet mount, on the other hand, is used when the specimen to be viewed is in water or some other liquid. To prepare a wet mount, use an eyedropper or similar equipment to place a few drops of a liquid sample onto the glass slide. Holding the coverslip by its edges, carefully place one edge of the coverslip on the slide near the sample. Gently lower the coverslip into place. Do not press down on the coverslip. This could damage the sample and possibly crack the coverslip or the slide. If liquid leaks out from under the coverslip, use a piece of absorbent paper to soak it up. Do not wipe the slide with the paper. This could cause the coverslip to slide and may damage the specimen.

To view a specimen that has been dry or wet mounted, place the microscope 2 to 3 inches (6 to 8 centimeters) from the edge of the workbench. When carrying a light microscope from one place to another, always carry it by the arm and the base. The arm of the microscope is the thick metal part that connects the lens tube to the stage. Always use two hands. Remove any dust from the microscope. Rotate the microscope's nosepiece until the lowest power objective (the shortest one) is in place. Use the coarse focus adjustment knob to raise the objective lens (or lower the stage, depending on the type of light microscope) so that the lens is in its highest position. The coarse focus adjustment is usually the larger of the two focus knobs. Open the diaphragm. Look through the ocular lens (the one in the eyepiece) and adjust the mirror or light source so light can be seen. Place the prepared slide on the microscope stage and secure it with the clips provided on the stage.

To focus on the specimen, use the coarse adjustment knob to slowly bring the objective lens closer to the stage. Be careful not to allow the lens to touch the coverslip at any time. When the specimen can best be seen (even if it is still a little blurry), switch to adjusting the fine focus knob. This will bring the sample into crisp focus so that the intricate details of the specimen can be seen.

If viewing the sample under a higher magnification is desired, begin by following the aforementioned steps. Some high-powered objective lenses require a layer of oil between the lens and the slide. This type of lens is called an **oil immersion lens**. If an oil immersion lens is being used, rotate the nosepiece until the low-power objective is out of the way and there is easy access to the slide. Place one drop of microscope oil on the top of the coverslip. Carefully rotate the high-power objective into place. The objective should not touch the glass coverslip. However, it will touch the drop of oil. If there is any resistance when trying to rotate the high-power

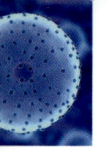

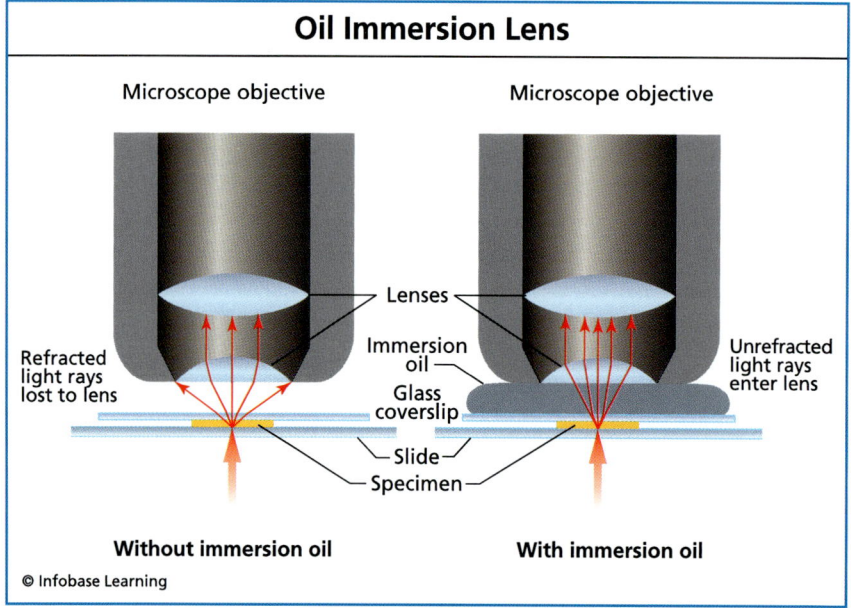

FIGURE 2.6 Typically, light rays (shown emanating from the light source and through the sample) are refracted as they leave the glass coverslip and enter the air. Refraction is minimized when oil is used to fill the space. The oil allows the lens to collect more light and improve image quality.

lens into place, rotate the objective away from the slide and go back to low power. Readjust the focus under low power and try again. Never force an objective lens into place. Damage to the lens or specimen can result. When the high-power objective is in place, use only the fine focus adjustment knob. Using the coarse focus adjustment can also result in damage to the slide or the lens.

ELECTRON MICROSCOPES

Light microscopes are limited in what they can reveal to scientists by the properties of light itself. Visible light travels in waves. Its **wavelength** is the distance between the crest of one wave and the crest of the next. The average wavelength of visible light is around 0.5 micrometers (2.0×10^{-5} inches). Any object smaller than this wavelength will not be seen, even with the most powerful light microscope.

Think of it this way: If someone tosses a rock into a pond, waves spread out over the surface of the water. Imagine that these waves have a wave-

PARAFFIN EMBEDDING AND MICROTOMES

Pathologists are doctors who specialize in examining bodily tissues for evidence of disease. Microscopic investigation of tissue requires that the specimen be thin enough for light to pass through it. Therefore, technicians must prepare pathology samples using special techniques.

Body tissues may be too thick to examine under a microscope, but they are still delicate. Attempting to cut them into thin pieces would shred them and microscopy would be impossible. Instead, the tissues are prepared by dehydrating (removing water from) them and injecting them with a waxy substance called paraffin. The paraffin-infused tissue is then placed into a box-shaped mold and melted paraffin is poured on top. When the paraffin cools and the mold is removed, the result is a block of wax that contains the body tissue to be studied. This block can be sliced into thin sections appropriate for microscope work without fear of damaging the tissue.

The process of slicing the tissue into thin pieces is called sectioning. Sectioning is done using a special instrument called a **microtome**. A microtome is similar to a meat slicer in a deli, but instead of moving back and forth over the specimen, the wheel of the microtome cuts the tissue in a circular motion. Sectioning is a very tricky process and requires skill. The tissue sections are very thin and easy to inadvertently tear, crease, or fold. Only perfectly flat, thin sections are useful for examination.

length of four inches (10.2 cm). When the waves encounter an object, like an eight-inch-long (20.3-cm-long) turtle, for example, they bounce off in different directions. These reflected and refracted waves could, theoretically, be analyzed and a rough picture of a turtle would emerge. However, if those waves encounter a tiny frog only two inches (5.0 cm) long, they will pass by undisturbed. The waves are not reflected or refracted and, therefore, the frog does not register. Likewise, a 0.2 micrometer virus (one of the larger viruses known) will not disrupt a light ray with a wavelength of 0.5 micrometer. Therefore, the virus cannot be seen.

Most cells, and even some of the larger structures inside cells, such as the nucleus, mitochondria, and chloroplast, can easily be seen under a

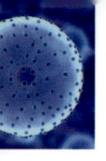

light microscope. Many types of bacteria can also be viewed under a light microscope. However, even the most powerful light microscope cannot resolve a virus. To see very small objects, such as a virus and the smaller cell organelles, a different type of microscope, called an electron microscope, is needed.

There are several different types of electron microscopes and they each work in a slightly different manner. Instead of using light to view an object, a transmission electron microscope (TEM) directs a high-voltage beam of electrons at the specimen. In addition, instead of glass lenses, a TEM uses electromagnetic lenses to focus the electrons into a very thin stream. Like the light waves from a light microscope, some of the electrons in the beam will be scattered when the electrons come into contact with the specimen. Other electrons are absorbed while others pass right through the sample. A fluorescent screen absorbs the electrons that pass through the sample and an image of the specimen is created. Scientists may study this image on the fluorescent screen itself, on a computer monitor, or photographs of the image may be taken for later study.

Specimens are prepared for viewing in a TEM by embedding them in hard plastic, similar to the process described for body tissues. An instrument called an ultramicrotome uses a diamond knife to slice the specimen into sections that are 0.05 to 0.10 micrometers (2.0×10^{-6} to 4.0×10^{-6} inches) thick. The TEM was introduced in the 1930s. The technology was co-invented by German scientists Max Knoll (1897–1969), an electrical engineer, and Ernst Ruska (1906–1988), a physicist.

Another type of electron microscope called the scanning electron microscope (SEM) appeared on the scientific stage in the 1950s. An SEM uses a set of coils to move the beam of electrons back and forth over the specimen, scanning it. As the beam moves across the sample, some electrons in the electron beam are scattered while others hit electrons within the sample and knock them out of place. High energy X-rays are also emitted along with the dislodged electrons. Detectors gather the X-rays and the dislodged and scattered electrons. The signal pattern created is used to produce an image of the specimen.

Another type of electron microscope, called the scanning tunneling microscope (STM), can produce a three-dimensional image of a specimen. In this type of electron microscope, a stylus with a pointed tip scans slowly back and forth over the sample at a fixed distance. The distance between the tip of the stylus and the surface of the specimen is tiny—only one atom wide. Electrons move, or tunnel, between the tip of the stylus and

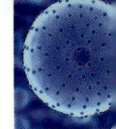

the sample. These electrons emit an electrical signal and the stylus moves up and down to keep that signal constant. These movements are incredibly small, just enough to maintain the distance of one atom between the tip of the stylus and the surface of the sample. The movement of the stylus is recorded, and a computer produces a contour map of the sample's surface.

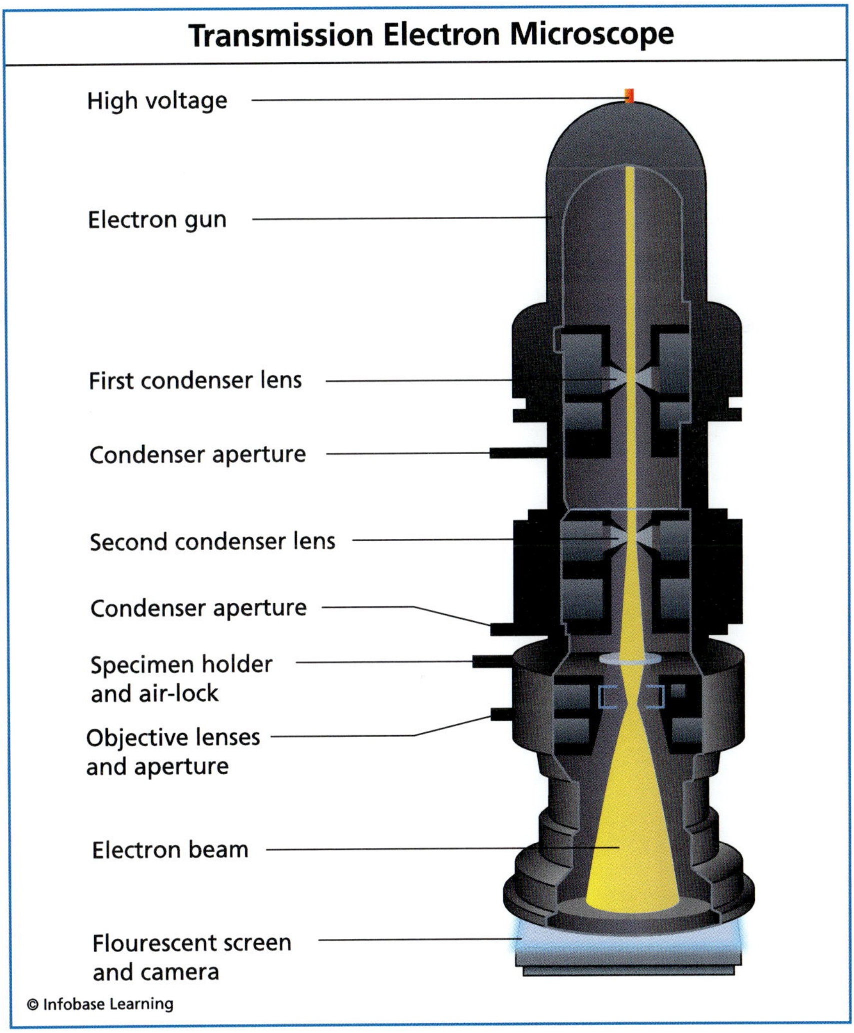

Transmission Electron Microscope

High voltage

Electron gun

First condenser lens

Condenser aperture

Second condenser lens

Condenser aperture

Specimen holder
and air-lock

Objective lenses
and aperture

Electron beam

Flourescent screen
and camera

© Infobase Learning

FIGURE 2.7 A transmission electron microscope features a beam of electrons that is transmitted through an ultrathin specimen. The interaction of the electrons transmitted through the specimen forms an image that is magnified and focused onto a screen, film, or other viewing device. These kinds of microscopes allow for significantly higher resolution than light microscopes.

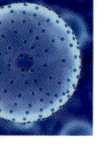

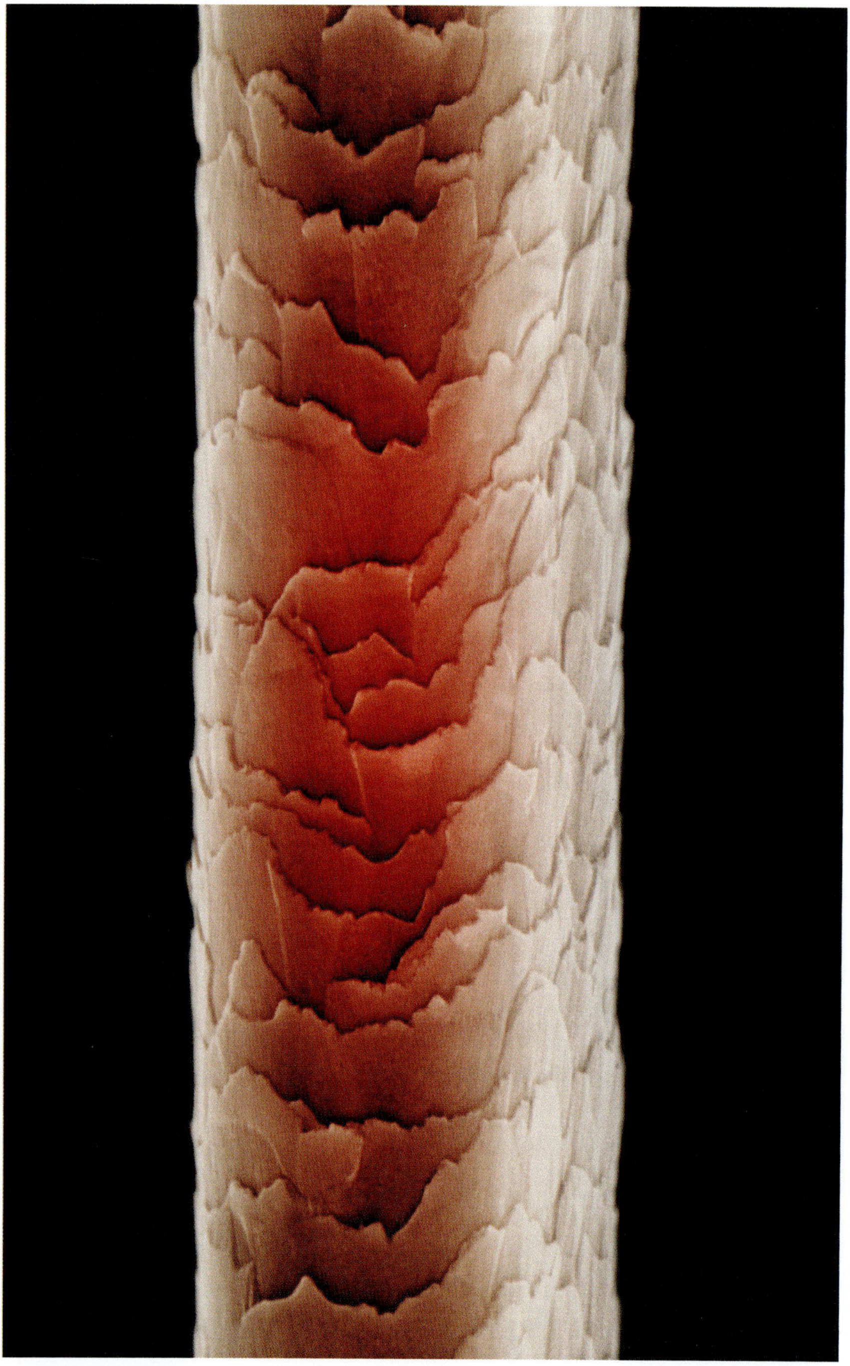

FIGURE 2.8 A human chest hair is magnified 500 times under a scanning electron microscope.

VIRUSES

Viruses do not have cells and they, themselves, are not cells. In fact, scientists do not even consider them to be living beings. They are tiny, infectious parasites that do not respire or grow. They do, however, reproduce: sometimes, at an alarming rate.

The first scientist to describe a virus was Russian botanist Dmitri Ivanovsky (1864–1920). In 1892, while studying diseased tobacco plants, he discovered that an agent capable of infecting healthy tobacco plants with the same disease would pass through ceramic filters that would otherwise capture even the smallest known bacteria. However, Ivanovsky refused to believe that any organism capable of causing disease could be smaller than bacteria.

In 1898, Dutch microbiologist Martinus Beijerinck (1851–1931) repeated and expanded on Ivanovsky's experiments. Unlike Ivanovsky, Beijerinck came to the conclusion that whatever was causing disease in the tobacco plants was indeed extremely tiny. Today, the virus studied by Ivanovsky and Beijerinck is called the tobacco mosaic virus.

In 1940, the advances in electron microscopy developed by Ernst Ruska allowed his brother Helmut Ruska (1908–1973) to take the first pictures of these super tiny parasites. Today, scientists continue to use the electron microscope and other techniques to identify and develop a better understanding of the nature of these nonliving parasitic particles.

Electron microscopes are still limited by the wavelength of the electron beam. However, this wavelength is much shorter than the wavelength of visible light (around 0.0002 to 0.0005 micrometers). Because of their shorter wavelengths, electron microscopes can reveal tiny structures in extraordinary detail. They are capable of magnifying objects up to 100,000 times their natural size. At this resolution, these microscopes can reveal structures as small as individual atoms.

3

Getting a Better Look: Staining Cells

Although many types of cells can be seen using a simple light microscope, they are not always that easy to distinguish. Most cells are transparent. Even if the faint outline of the cell membrane can be resolved, the organelles can be extremely hard, if not impossible, to see. To alleviate this problem, researchers have developed a number of staining techniques to color the transparent parts of cells and organelles so that they can be seen more clearly.

HISTORY OF CELL STAINING

In 1856, English chemist William Henry Perkin (1838–1907) developed the first synthetic dye, mauveine, when he was 18. At the time, Perkin was attending London's Royal College of Chemistry. On the advice of his teacher, August Wilhelm von Hoffman (1818–1892), Perkin began looking for a way to make quinine, an antimalarial drug. At the time, quinine could only be obtained from the bark of the cinchona tree. These trees only grew in the jungles of South America and on plantations in Southeast Asia. While Perkin never found a way to make synthetic quinine (this was accomplished by two American chemists, Robert Burns Woodward and William von Eggers Doering, in 1944), he did launch the chemical dye industry, and, indirectly, the pharmaceutical industry. In 1857, Perkin opened a factory to produce and sell his newly discovered dye and to discover other chemical dyes.

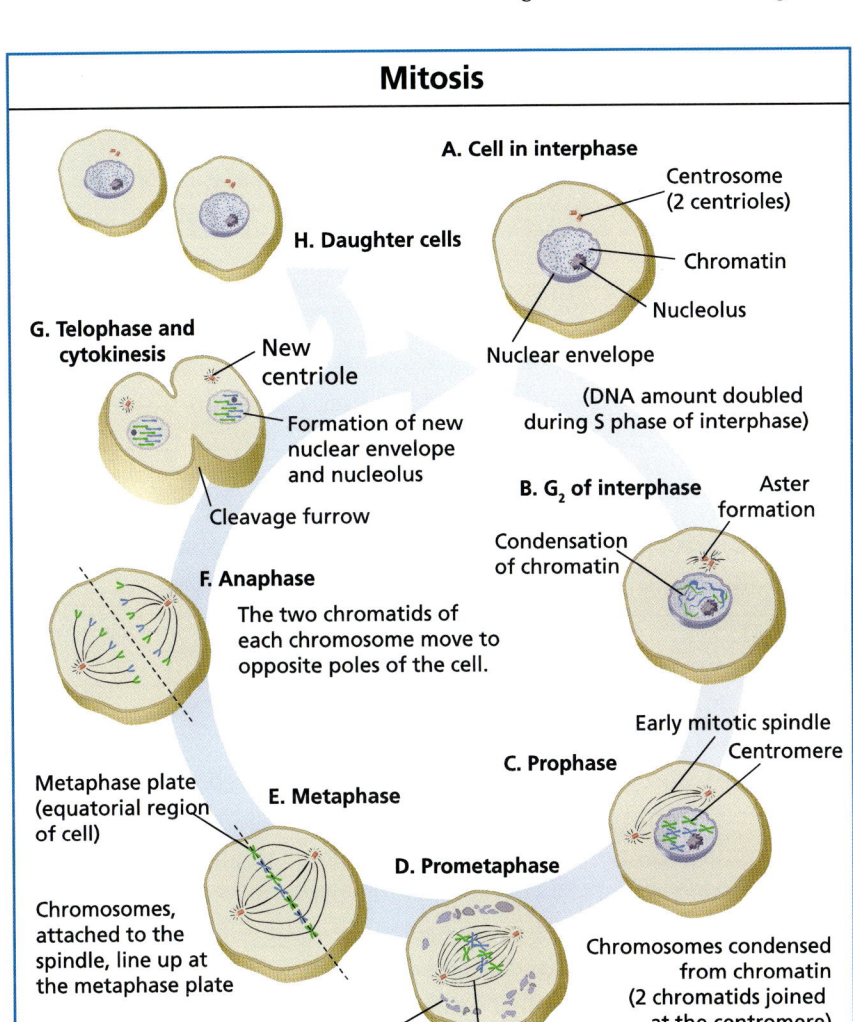

Mitosis

A. Cell in interphase

Centrosome (2 centrioles)

Chromatin

Nucleolus

H. Daughter cells

Nuclear envelope

(DNA amount doubled during S phase of interphase)

G. Telophase and cytokinesis

New centriole

Formation of new nuclear envelope and nucleolus

Cleavage furrow

B. G₂ of interphase   Aster formation

Condensation of chromatin

F. Anaphase

The two chromatids of each chromosome move to opposite poles of the cell.

Early mitotic spindle

Centromere

C. Prophase

Metaphase plate (equatorial region of cell)

E. Metaphase

D. Prometaphase

Chromosomes, attached to the spindle, line up at the metaphase plate

Chromosomes condensed from chromatin (2 chromatids joined at the centromere)

Nuclear membrane and nucleolus disintegrate   Spindle

© Infobase Learning

FIGURE 3.1 Mitosis is a multistage process. (A) During interphase, a cell carries out its normal functions. (B) In the second gap phase (G2), a cell completes replication of its centrioles and synthesizes enzymes that control cell division. (C) Chromatin condenses into chromosomes, spindle fibers elongate, and nucleoli and the nuclear envelope disappear during prophase. (D) Some of the spindle fibers attach to chromosomes, causing them to move during prometaphase. (E) Metaphase is characterized by chromosomes lining up along the midline of a cell. (F) This is followed by anaphase, where spindle fibers pull sister chromatids to opposite poles of a cell. (G) In telophase, chromatin decondenses, a new nuclear envelope and nucleoli appear, and spindles vanish. (H) Completion of this cycle results in two daughter cells with identical sets of genes.

Perkin's discovery led to the opening of a number of dye-making factories in England and Germany. Between them, these factories produced and tested thousands of different textile dyes in a rainbow of colors. However, the textile industry was not the only entity to make use of these dyes. Scientists would also eventually use them to make biological tissues and cells more visible.

In fact, it was Perkin's aniline dye that Walther Flemming used in 1879 to identify chromosomes and to observe and describe mitosis in animal cells. The fact that the red dye was absorbed so well by the chromosomes helped give the structures their name. *Chromatin* comes from a Greek word for "color." Because of his discoveries, Flemming is often cited as the father of the cytogenetics. Cytogeneticists study the structure and function of chromosomes.

Flemming's observations of cell division were not easy to make. Microscopes at the time did not have very good resolution (the ability to make objects appear clearly). In order to see the chromosomes and determine how they moved during mitosis, Flemming had to meticulously prepare and stain numerous microscopic samples at different stages of cell division. Scientists would require several more decades and advances in light microscopy before they could confirm Flemming's findings by observing the division of a live cell.

Flemming was not the first scientist to apply the textile dyes to biological specimens. In 1871, German pathologist Karl Weigert (1845–1904) discovered that different strains of bacteria reacted differently to particular dyes. Because they stain distinctly, different types of bacteria can be distinguished from one another using biological stains. However, Weigert is probably better remembered for his method of staining myelin sheaths, the insulating layer that surrounds the axons of nerve cells. Weigert's staining method helped scientists determine the anatomy of the central nervous system.

In 1880, Weigert discussed the need for dyes that would stain cells consistently with a graduate student named George Grubler. The textile dyes that scientists had been using up to this point did not always produce uniform results. Differences in composition of a biological stain can lead to improper staining of a specimen. Improper staining can make a specimen hard to see and impurities in the stain can produce artifacts. Shortly after his graduation, Grubler founded a company to provide scientists with biological stains. Grubler sold his stains to scientists from all over the world, and his dyes soon became known as some of the best in the

scientific community. In fact, some staining protocols specifically named Grubler's dyes as the ones to use.

One of the stains that Grubler's company produced and distributed was crystal violet (called gentian violet at the time). In 1882, Paul Ehrlich (1854–1915), a renowned immunologist and Karl Weigert's cousin, developed a staining method that would reliably stain the bacteria that caused tuberculosis. That same year, Hans Christian Gram (1853–1938), a Danish bacteriologist, modified Ehrlich's staining method to differentially stain bacteria in a way that aided in their identification. (Gram published his method in 1884.) Still in use today in hospitals and research laboratories around the world, Gram staining is considered to be one of the most important staining techniques in microbiology.

Grubler stains continued to be the most-used biological stains worldwide until sometime during the years 1916 to 1918. During World War I (1914–1918), German products, including biological stains from Grubler's company, became unavailable to scientists in the United States and in the United Kingdom. This situation required American and British scientists to seek out other sources for their stains. By 1920, the war was over, but American scientists were still unable to obtain German stains. That year, the American scientists discovered that the stains they had been using were not uniform and, therefore, did not always provide reliable results. To combat this problem, the National Research Council and several scientific societies worked together to form the Commission on Standardization of Biological Stains in 1921. This nonprofit organization is now known as the Biological Stain Commission. Today, biologists and chemists associated with this commission study and gather information about the nature of dyes and how they work in microscopic samples. In addition, the commission works with manufacturers who produce biological stains to make sure that the stains used in scientific laboratories are of the highest quality. The commission's laboratories are now located at the University of Rochester Medical Center in Rochester, New York.

HOW ARE CELLS STAINED?

Different types of stains and different staining techniques are used to color different parts of the cell. One stain, for example, might be used to allow a scientist to get a good look at the nucleus while a different stain might be used to visualize the cell membrane.

Some stains can only be used on nonliving samples. Others can be used on either living or dead cells. The staining technique will differ

A CHEEKY EXPERIMENT

Many of the stains used in research laboratories are sophisticated and hard to find, but not all of them. Some simple biological stains include food coloring, iodine, malachite green, and methylene blue. Food coloring can be found in the grocery store, iodine in the pharmacy, and malachite green and methylene blue can be obtained from an aquarium store. Here is an easy experiment that will show how staining cells helps scientists see them better:

Materials needed:

- toothpick
- two microscope slides
- eyedropper
- water
- two coverslips
- stain (food coloring, iodine, malachite green, and/or methylene blue)
- paper towels
- microscope

depending on the type of sample being used and on the type of stain. In some cases, for example, a sample might be prepared for staining by adding a chemical called a surfactant. The surfactant dissolves the cell's membrane and allows larger stain molecules inside the cell. Because it allows the stain to pass through the cell membrane, this step is called permeabilization (derived from the word *permeable*).

In some staining techniques, a step called fixation might be used, too. Fixing cells preserves their structure during the staining process. This step normally involves adding a chemical to the sample that creates chemical bonds between proteins. These bonds help preserve the cell's shape. Some of the chemicals used to fix cells include formaldehyde, ethanol, and methanol.

Once the cells are prepared, they are mounted on microscope slides. Mounting cells on slides can be done in several different ways. For example, the cells may be grown directly on the slides. With this approach,

Gently scrape the flat edge of a toothpick along the inside lining of your cheek to pick up some cheek cells. Wipe the flat edge of the toothpick onto one of the clean microscope slide to transfer the cheek cells. Use the eyedropper to place one drop of water on top of the cheek cells. Rest one side of a coverslip on the slide, next to the drop of water, and carefully lower the other side so that it covers the cheek cells and the drop of water. Repeat this procedure with the other slide.

Now that you have prepared two cheek cell samples, you need to stain one of them. To add stain to one of the slides, place one drop of stain next to one edge of the coverslip. Place a paper towel on the other side of the coverslip. The paper towel will draw, or wick, the stain under the coverslip. Look at the stained and unstained slides under a microscope. Can you see the cells better on the stained slide? Repeat the experiment as often as you would like using different stains.

CAUTION: Cells are not the only things that can be colored with stains. Hands, clothing, fabric, and other materials can, too. To protect the area you are working in, you may want to cover it with layers of paper towels or newspapers. To protect your clothing, an old apron would be appropriate and useful.

permeabilization and fixation occur while the cells are still on the slide. Individual cells, like the cheek cells, can also be applied to a slide after permeabilization and fixation. Tissues, or groups of cells, can also be placed on a microscope slide and prepared for staining.

Cells can be stained before or after they are fixed and mounted on slides. Staining often involves dipping the cells into a liquid dye. Other stains require the use of another chemical called a mordant. When a mordant is added to the slide, a chemical reaction occurs between the stain and the mordant. The chemical reaction produces a colored precipitate. (A precipitate is an insoluble solid, meaning that it does not dissolve). Sometimes a counterstain may also be used to increase contrast or to further distinguish certain cell types or structures.

After the slides are stained, they need to be kept in a dark, cool spot to preserve the staining. Many laboratories place their slides in slide boxes inside a refrigerator to store them long term.

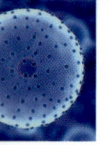

FIGURE 3.2 The materials needed to make a Gram stain—typically the first step in the identification of bacterial organisms—are shown.

GRAM STAINING

Gram staining, the staining method developed by Hans Christian Gram in 1882, is still in use today and is usually the first test to be carried out in order to identify bacteria. Bacteria can be separated into two groups—Gram-positive bacteria and Gram-negative bacteria.

Gram staining is done in three steps—staining with a primary stain, decolorizing, and counterstaining. The primary stain in Gram staining is crystal violet. However, sometimes, methylene blue is substituted for the crystal violet and this works as well. Next, an iodine solution, called Gram's iodine solution, which is a mixture of iodine and potassium iodide, is added. The iodine is a mordant. The iodine solution and the crystal violet chemically react to form a precipitate. This solid cannot be dissolved in water. This step is often called fixing and it means securing the dye into place.

Next, a decolorizer, usually ethyl alcohol or acetone, is added to the sample. The exact method for how the decolorizer works is not well understood and is still being debated in the scientific community. However, one explanation proposes that the decolorizer dehydrates the lipid layer in the cell membrane. This dehydration causes the lipid layer to tighten and shrink. The crystal violet-iodine precipitate is a relatively large molecule and it cannot pass from the interior of a Gram-positive cell through the tightened lipid layer. The membrane surrounding Gram-negative bacteria, however, is much thinner. In fact, Gram-positive bacteria might have as much as twenty times more lipid molecules in their cell membrane as a Gram-negative bacteria does. Because the crystal violet-iodine complex is retained in the Gram-positive bacteria, these bacteria appear to be a deep

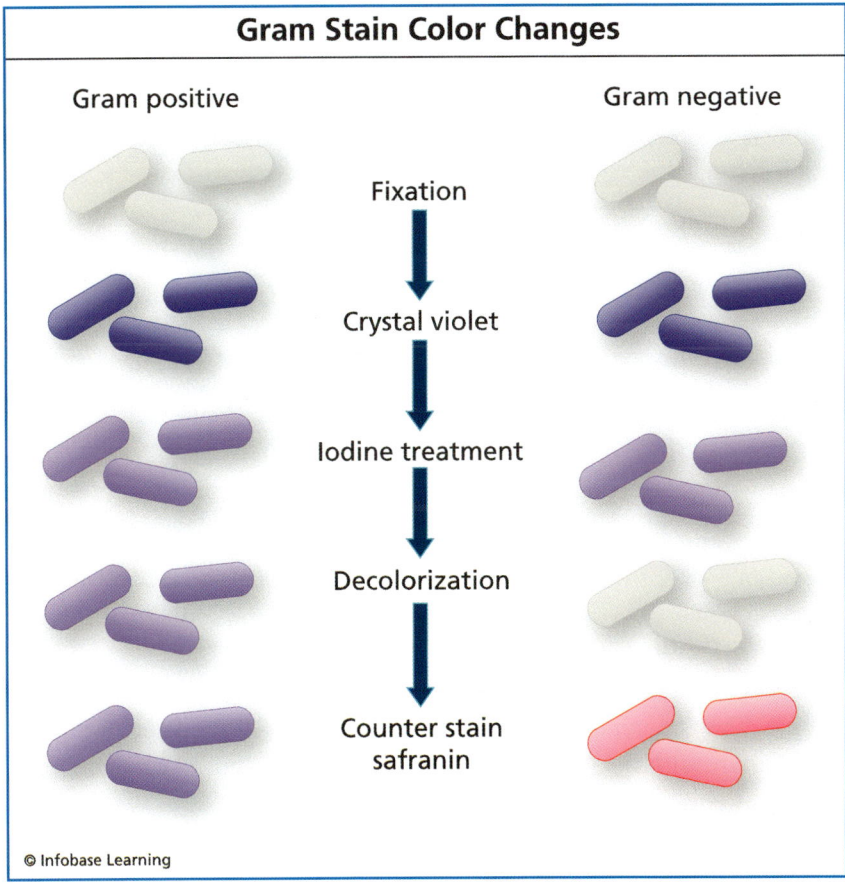

Gram Stain Color Changes

Gram positive

Gram negative

Fixation

Crystal violet

Iodine treatment

Decolorization

Counter stain safranin

© Infobase Learning

FIGURE 3.3 The Gram stain enables most bacteria to be divided into two groups: Gram positive and Gram negative. This technique is based on the idea that a Gram-positive cell wall has a stronger attraction for crystal violet than a Gram-negative cell wall when Gram's iodine is applied.

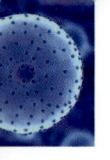

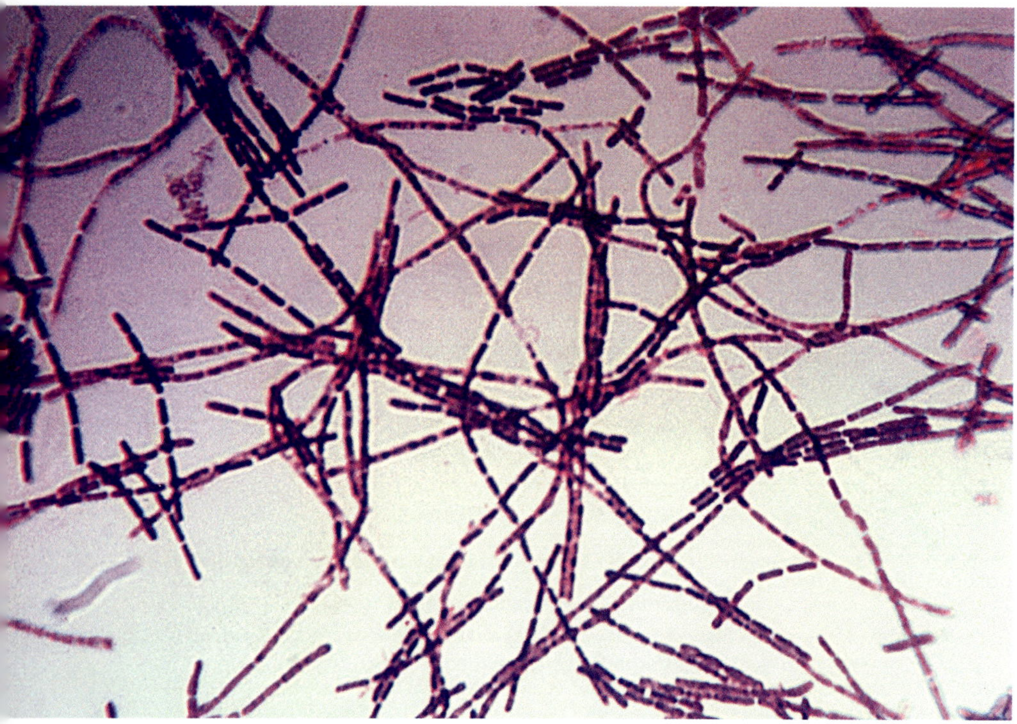

FIGURE 3.4 This photomicrograph shows *Bacillus anthracis* bacteria using Gram-stain technique. Anthrax is diagnosed by isolating *B. anthracis* from blood, skin lesions, or respiratory secretions, or by measuring specific antibodies in the blood of people with suspected cases of anthrax.

purple under the microscope. However, in Gram-negative bacteria, which started out with a thinner membrane, the decolorizer dissolves the lipid layer. No longer able to retain the crystal violet-iodine complex inside the cell, the Gram-negative bacteria loses its purple color.

Finally, a counterstain, usually a stain called safranin, is added to the sample. The counterstain dyes the bacteria pink. The counterstain does not interfere with the deep purple color of the Gram-positive bacteria, but it allows the decolorized Gram-negative bacteria to be seen easily under the microscope.

Streptococcus pyogenes (the bacterium that causes strep throat), *Clostridium tetani* (which causes tetanus), and *Bacillus subtilis* (a soil-dwelling bacteria) are all examples of Gram-positive bacteria. *Escherichia coli* (a helpful bacteria when it stays in the digestive tract where it

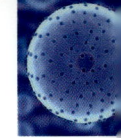

belongs, but can cause severe food poisoning when it does not), *Neisseria meningitidis* (the bacterium that causes meningitis), and *Borrelia burgdorferi* (the bacterium that causes Lyme disease) are examples of Gramnegative bacteria.

OTHER TYPES OF STAINS

Different stains color different parts of the cell. Scientists can also use different types of stains to distinguish between live cells and dead ones, to highlight certain metabolic processes, and to detect pathogenic microorganisms, such as bacteria, in medical samples.

Methylene blue is a stain that is used to make the cell nucleus more visible. It is also a component of Wright's stain, a stain developed by American pathologist James Homer Wright (1869–1928) in 1902. Wright's stain is used to distinguish between different types of blood cells. The stain may also be used in combination with other stains. Wright-Giemsa stain, for example, is used to stain chromosomes for cytogenetic analysis. Cytogenetic analysis is used to determine if particular chromosomal abnormalities are present.

Methylene blue can also be used as a counterstain in Ziehl-Neelsen staining. This staining method, also called acid-fast staining, is used to determine if human sputum or nasal scrapings contain the mycobacteria that cause tuberculosis. Mycobacteria have unusually thick, waxy cell walls and can be hard to stain with Gram stain.

Nile blue and toluylene red are other stains that can be used to make cell nuclei more visible. Both of these stains may be used on living cells.

Iodine is used by biologists as an indicator for starch. If an iodine solution is added to a starch, a starch-iodine complex forms. This starch-iodine complex creates a deep blue color. Iodine is also a component of Lugol's solution, which is used as an antiseptic (a substance that inhibits the growth of microorganisms, such as bacteria). First made in 1829 by French physician Jean Guillaume Auguste Lugol (1786–1851), Lugol's solution is a mixture of pure iodine and potassium iodide in water. Lugol's solution turns black in the presence of starch. The solution may also be used as a cell stain to make the nucleus more visible and as a substitute for Gram's iodine solution in Gram staining.

Scientists use many other types of stains, too. For example, DAPI (4',6-diamidino-2-phenylindole) is a fluorescent stain that binds tightly to DNA in the nucleus and glows a vivid blue when excited by ultraviolet light. This stain helped scientists to determine that mitochondria contained its own DNA.

Another fluorescent stain, ethidium bromide (EB), is often used with another stain, acridine orange (AO) to identify healthy and unhealthy cells. The EB-AO combination stains healthy cells fluorescent green, while unhealthy cells in the last stages of cell death (apoptosis) are dyed fluorescent red-orange.

WHO ARE THESE SCIENTISTS?

Many different types of scientists study cells. The names of their specialties can be confusing, but they do a good job at describing the types of cells and in what situations scientists study them. Here is a list of scientists who study cells (and other related organisms) and what their particular specialties mean:

- **bacteriologist** A biologist who studies bacteria
- **biochemist** Studies the chemical makeup of living things
- **cell physiologist** Studies cell function and how cells interact with their surroundings
- **cytochemist** A biochemist who specializes in studying the chemical makeup of cells
- **cytogeneticist** A type of geneticist who studies the hereditary components of a cell
- **cytologist** A biologist who studies the structure and function of cells
- **histologist** A specialist who studies how cells are organized and how they function in tissues
- **microbiologist** A scientist who studies microscopic organisms, including bacteria, viruses, fungi, and protists
- **molecular biologist** A biologist who studies life at the molecular level
- **mycologist** A botanist who specializes in fungi
- **pathologist** A scientist (or doctor) who specializes in the examination of cells, tissues, and organs in order to arrive at a medical diagnosis of disease
- **virologist** A biologist who studies viruses

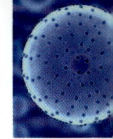

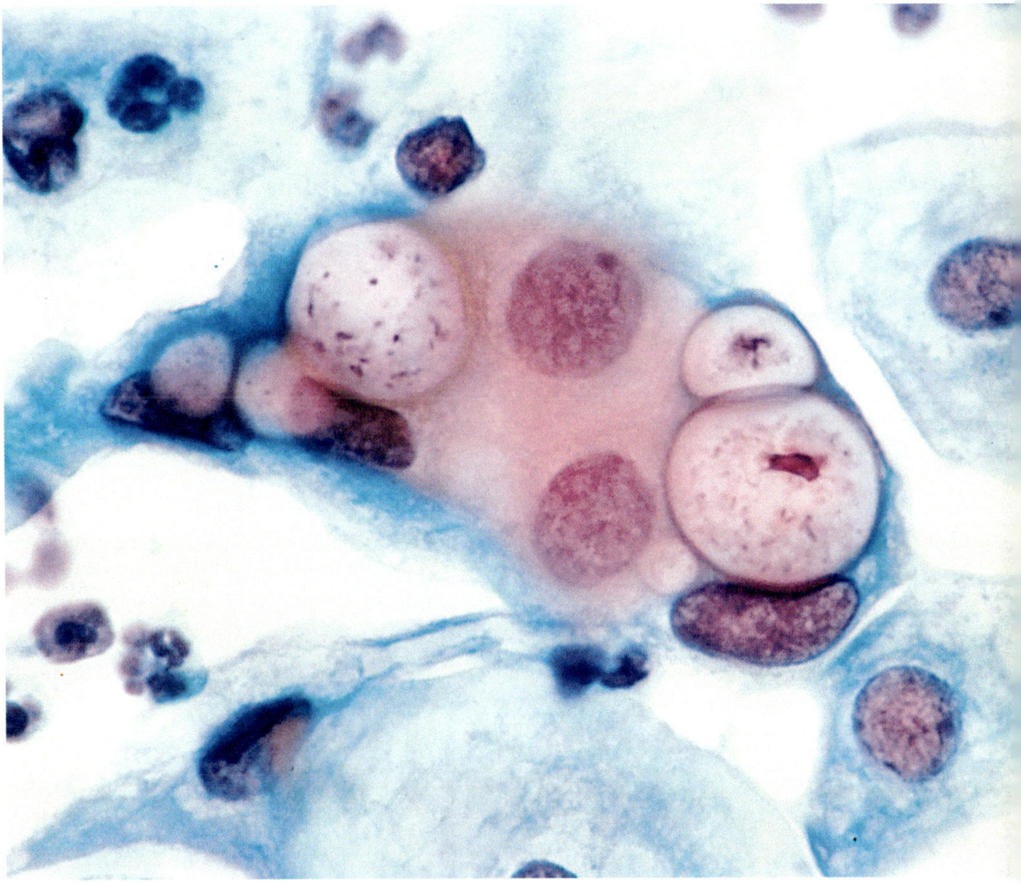

FIGURE 3.5 With a hematoxylin-eosin stain, a human Pap smear shows the presence of chlamydia in the vacuoles at a 500x magnification.

Rhodamine, another fluorescent dye, binds to proteins. This dye may be used to determine how proteins move in and out of cells. Because rhodamine is a fluorescent dye, its movements can be detected with the use of a device called a fluorometer.

Osmium tetroxide stains lipids black and can be used in optical as well as electron microscopy. Hematoxylin-eosin stain (also called H&E staining) dyes the cytoplasm bright red and the nuclei dark blue. H&E staining is one of the most common staining procedures used in histopathology (the microscopic examination of abnormal tissue). These are just a few of the thousands of staining procedures used in biological laboratories around the world every day.

4

Cell Culture

To study cells, scientists sometimes need to grow them in the laboratory, independent of a living organism. The general term for growing and maintaining cells, tissues, and organs in the laboratory is *tissue culture*. Procedures performed in a controlled environment, such as a scientific laboratory, outside a living organism are called **in vitro** procedures. The term *in vitro* comes from the Latin for "within glass." Procedures performed on a living organism, on the other hand, are called **in vivo**. *In vivo* also comes from Latin and means "within the living." In order for cells to live and divide outside a living being, they must be provided with the correct nutrients and physical conditions.

HISTORY OF CELL CULTURE

The first scientist to overcome the problems of growing cells outside of a living organism was American biologist Ross Harrison (1870–1959). While working at The Johns Hopkins University, Harrison was trying to learn how nerve cells grew in embryos. Other scientists had studied nerve cells in the laboratory before, but none of these scientists had tried growing and experimenting on them. In 1907, Harrison published a paper detailing his successful attempt to grow frog nerve cells in the laboratory for the first time. To do this, he placed small pieces of a frog embryo's neural tube and a drop of the fluid (lymph) that is found between cells in the body onto a coverslip. When the lymph clotted, Harrison turned the

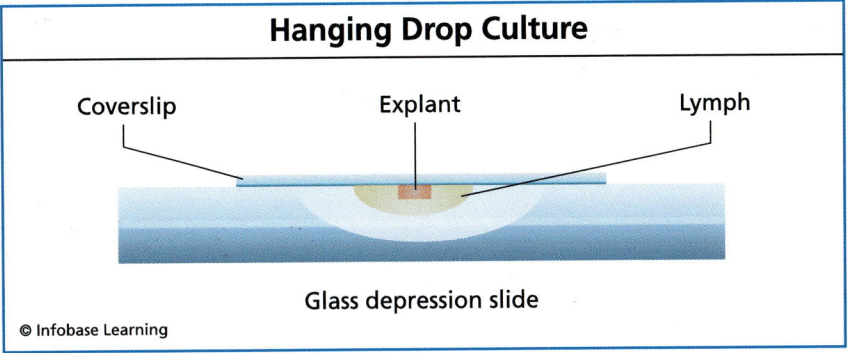

Hanging Drop Culture

Coverslip   Explant   Lymph

Glass depression slide

© Infobase Learning

FIGURE 4.1 A hanging drop culture provides a means for allowing growth that would otherwise be restricted by the flat plane of petri dishes. It also minimizes the surface area to volume ratio, slowing evaporation.

coverslip upside down and put it on top of a microscope slide that had been made with a depression (also called a well). This tissue culture technique is called a hanging drop culture. Harrison continued his research at Yale University and from 1907 to 1910 published several papers detailing his new techniques. Today, microbiologists still use Harrison's hanging drop culture technique to grow bacteria in the laboratory.

The year before Harrison's breakthrough paper on frog nerve cells was published, French surgeon and biologist Alexis Carrel (1873–1944) arrived at the Rockefeller Institute for Medical Research (now called Rockefeller University). At that time, Carrel was already a celebrated surgeon. He had developed a technique that allowed blood vessels to be sutured (sewn) together without the formation of blood clots. This technique allowed Carrel to perform the first organ transplants in animals. It also sparked Carrel's interest in growing replacement organs outside of the body. In 1910, Montrose Burrows (1884–1947), one of Carrel's colleagues, traveled to Yale to observe and learn about Harrison's tissue culture techniques. By the end of 1910, with Burrows's help, Carrel had successfully cultured chicken, rat, dog, and human tumor cells outside of the body. This team of scientists also learned how to split, or **subculture**, some of their cultures by carefully cutting them up and transferring them to another culture vessel. These cells were the first cell lines. A cell line is a group of cells that originate from one cell, are grown in vitro, and (with the exception of spontaneous mutations) are genetically identical. The ability to subculture their cell lines meant that Carrel and his team could keep the samples alive and growing for several months.

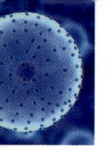

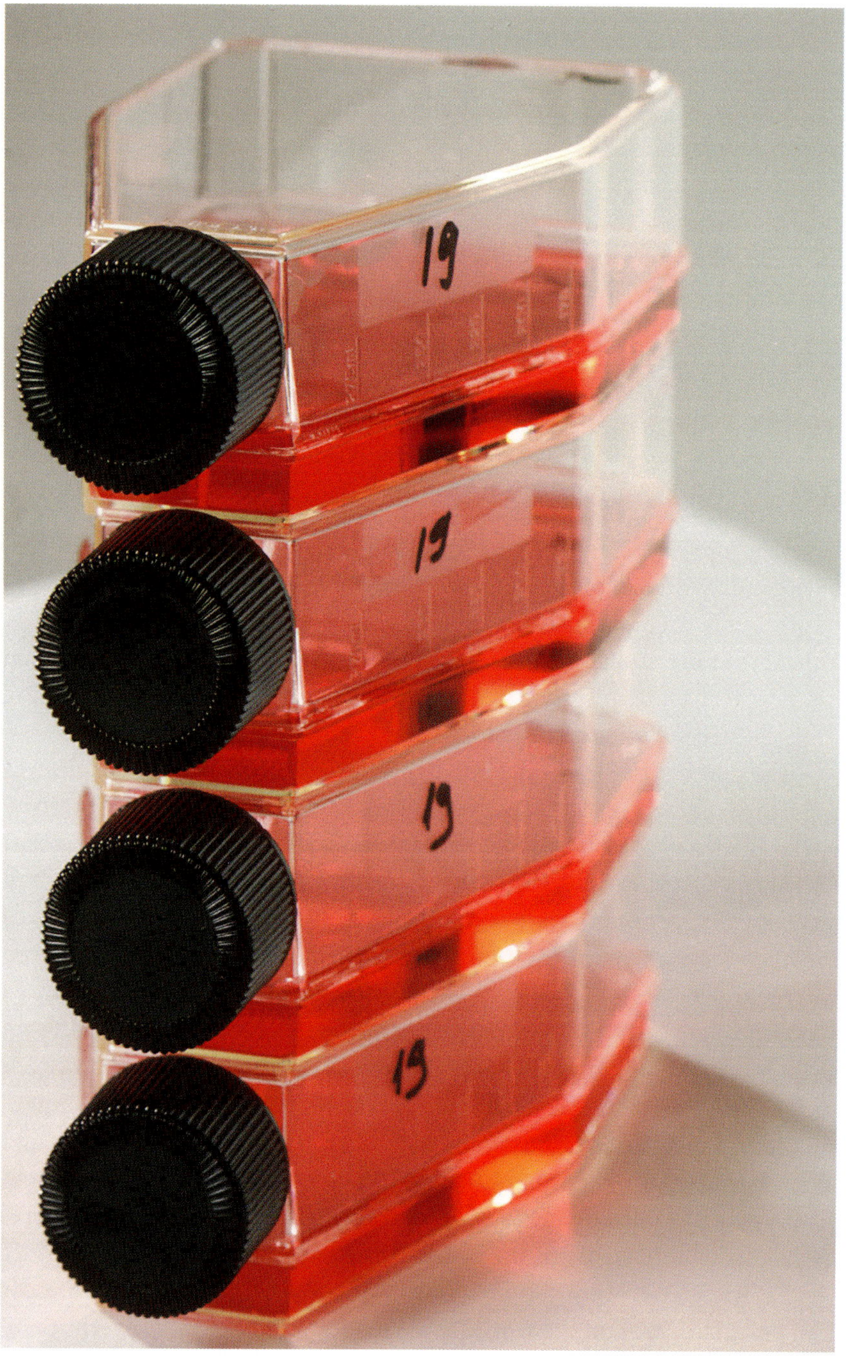

FIGURE 4.2 These vials contain tissue culture medium, which provides nourishment to growing cells.

Carrel soon found that if an extract of chick embryo was added to the cell culture's **growth medium** (the liquid the cells are grown in), the cell lines could be maintained indefinitely. In 1912, Carrel established a culture from fragments of a chick embryo's heart. As a former surgeon, Carrel had developed a rigorous set of laboratory rules to prevent his cultures from becoming contaminated with bacteria or fungi. (Contamination leads to cell death.) Carrel was working during a time before antibiotics were available to treat contaminated cell cultures. His success in maintaining cell lines relied mainly on his fastidious sterile technique. Later, another researcher who worked with Carrel, Albert Ebeling, took over the cell line. By following Carrel's strict sterile, or aseptic, technique, Ebeling successfully maintained the chick embryo cell line until 1946, two years after Carrel's death. By the time Ebeling terminated the cell line, it had been subcultured hundreds of times.

Another innovation that came out of Carrel's lab was the development of the cell culture flask. Up to this point, Carrel and other scientists had been following Harrison's hanging drop technique and suspending cells in a drop of plasma. However, the hanging drop cultures did not contain much growth media and the cells quickly used up all the nutrients available to them. To keep the cells alive, scientists had to split the cells every two to three days and transfer them to fresh growth media. In 1923, Carrel developed the D-3.5 culture flask. At 3.5 centimeters (1.3 inches) in diameter, Carrel's flask could hold more growth medium than other flasks. In addition, growing the cells in a flask allowed the cultures to be refed. (**Refeeding** a culture means to replace the old, depleted culture medium with fresh, nutrient-rich medium.) This breakthrough meant that cultures could be maintained for weeks before the density of cells required subculturing. By the mid-1920s, nearly every research laboratory in the world was using the culturing techniques developed in Carrel's laboratory.

Charles A. Lindbergh (1902–1974) is, no doubt, remembered more for his solo cross-Atlantic airplane flight from New York to Paris in 1927 than for his pioneering work in the tissue culture laboratory. What is not so well known is that Lindbergh spent much of the decade after his famous flight in Alexis Carrel's laboratory. With Carrel as his mentor, Lindbergh used his knowledge of physics to invent several devices that aided scientists in their studies of tissue and organ culture. The most famous of Lindbergh's devices is a pump used in organ culture. This glass chambered vessel pumped blood through the cultured organ to keep it alive and functioning. Often called an artificial heart by the press, Lindbergh's

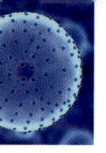

FIGURE 4.3 Famed pilot Charles Lindbergh (*left*) French surgeon Alexis Carrel (*right*) stand by their invention, a perfusion pump, which keeps human organs alive outside the body during surgery, in a photo taken in 1938.

invention landed him and Carrel on the cover of *Time* magazine in 1935. In 1939, Lindbergh and Carrel jointly published a book titled *The Culture of Organs*. In the 1950s, their work was used as the foundation for the development of the first heart-lung machines, which made open-heart surgery a possibility for the first time. Lindbergh also designed a cell culture flask that allowed the nutrient-rich liquid that the cells grow in to be continuously circulated, as well as other devices used in the laboratory.

Tissue culture has come a long way since Ross Harrison's time. Today, pharmaceutical companies culture certain bacteria and other cells by the billions. Bioreactors, some as large as 5,283 gallons (20,000 liters), contain cells that produce, through their metabolic processes, medications that are worth billions of dollars a year. The techniques that allow these huge bioreactors to exist today were largely developed in the 1950s. In 1953, cells were for the first time grown suspended, or floating, in culture medium instead of being grown on a surface, such as a slide or a coverslip. This culturing method is called a **suspension culture**. Suspension cultures are a common way to culture microorganisms. With the exception of blood cells, it is a less common way to grow animal or human cells. Most mammalian cells grow better attached to a surface. This type of culture is called an **adherent culture**. By 1957, suspension culture techniques approached volumes that would make the technique commercially useful for producing cell-based drugs.

WHY CULTURE CELLS?

Scientists need to grow cells in culture for many different reasons, including studying the biochemical processes in cells, testing different chemicals (such as drugs) on different cells types, and studying the possibility of generating artificial tissues. Over the last decade, the increase in the number of applications for growing genetically modified microorganisms that are capable of producing useful therapeutic chemicals has also made these techniques invaluable.

One of the first of these applications, discovered in 1948, was the ability to grow the poliovirus in tissue culture. There are two different types of polio vaccine, both of which are made from the virus that causes polio. One type of vaccine is made from dead poliovirus. This vaccine was discovered by American virologist Jonas Salk (1914–1995) in 1952. The other is made from the live, but weakened, virus. This type of vaccine was developed by Polish virologist Hilary Koprowski (1916–) and Polish-American researcher Albert Sabin (1906–1993).

For research purposes and to produce the vaccine, Salk and Sabin needed to discover a way to grow large amounts of poliovirus. Scientists knew how to grow some types of viruses. The influenza virus, for example, had been grown in chicken eggs in order to provide viruses for a flu vaccine. However, the poliovirus would not grow in eggs. In 1936, Sabin and Peter Olitsky (1886–1964) found that they could grow the poliovirus in cultures of human embryonic brain cells. However, scientists were concerned that a virus grown in brain cells might affect the central nervous system more often than a virus grown in other types of cells. In 1948, the

HELA CELL LINE: THE CELLS OF HENRIETTA LACKS

When normal cells are taken out of the body and grown in the laboratory, they go through a certain number of cell divisions before they begin to die. Some cancer cells, however, can live on indefinitely. One such immortal cell line, called HeLa, originated from the cancerous cervical cells obtained from a 31-year-old, African-American woman named Henrietta Lacks (1920–1951). Lacks never learned that some of her cells had been put aside for research, but since 1951, HeLa cells have been grown in laboratories all over the world.

Since that time, scientists have generated more than 60,000 scientific papers describing their research on the HeLa cell line. Over the years, HeLa cells have been used to research the polio, mumps, measles, and herpes viruses. They have helped scientists develop improved tissue culture techniques and equipment. New cryogenic methods that can keep cells alive while being shipped through the mail have been established using HeLa cells. In 1953, scientists working with the cell line discovered an improved staining method that would reveal the structure of human chromosomes. This development eventually led to the discovery that Down syndrome (a genetic disorder) was the result of an extra copy of chromosome 21. Other genetic breakthroughs followed. In 1954, scientists experimenting with HeLa cells discovered a way to isolate a single cell and how to keep that single cell alive long enough so that it can make an exact copy of itself. This breakthrough laid the foundation for gene therapy, in vitro fertilization, and animal cloning. In 1960, HeLa cells were sent into space. Scientists discovered that cancer cells grow faster in space. In 1984, a German virologist used HeLa cells to prove that the human papillomavirus causes cervical cancer (the disease that killed Lacks 33 years before),

American virologists John Franklin Enders (1897–1985), Thomas Huckle Weller (1915–2008), and Frederick Chapman Robbins (1916–2003) successfully showed that the poliovirus would infect non-nervous system cells grown in culture, too. This breakthrough paved the way for large-scale production of the virus and the vaccine. In culture, the poliovirus

which led to the development of the HPV vaccine.

This is just a partial list of the research advances that scientists have been able to make using the HeLa cell line. Meanwhile, research continues. More recently, scientists have discovered that HeLa cells produce an enzyme called telomerase. This enzyme allows cancerous cells to repair DNA damage that would kill a normal cell. Cancer drugs that target this enzyme are currently in the early stages of clinical trials. Henrietta Lacks may not have consented to have her cells used in this manner, but she has, no doubt, contributed much to medical science, and her legacy lives on.

FIGURE 4.4 In 1951, Henrietta Lacks became the unwitting donor of cells from a cancerous tumor, which were cultured to form a cell line—known as the HeLa cell line—for medical research. HeLa cells make up the first human cell line. Lacks died of cervical cancer eight months later, but HeLa cells are now used in cancer research worldwide.

multiplies six to eight hours after entering a cell, and 10,000 to 100,000 virus particles are produced per infected cell. Enders, Weller, and Robbins were awarded the 1954 Nobel Prize in Physiology or Medicine for this discovery.

Cells grown in culture normally contain only one type of cell. The cells may be genetically identical or they may show some genetic diversity. A culture of genetically identical cells is called a homogenous population. A culture that contains cells that have genetic differences is called a heterogeneous population. Homogenous populations that originate from one parental cell (due to cell division) are called **clones**.

Normal cell types have a finite life span and will die a natural death after a certain number of cell divisions. Other types of cells, such as cancerous cells, for example, can grow indefinitely in a properly maintained cell culture. Cells that can be grown indefinitely in tissue culture are called **continuous cell lines**. Sometimes in vitro cells start out as normal but, through contact with viruses, radiation, chemicals, or oncogenes, they become transformed. **Transformed cells** are immortal and, once converted, can reproduce indefinitely if maintained.

MAINTAINING CELLS IN CULTURE

To grow cells in culture, they must be provided with the proper nutrients. Cells are grown in a liquid called growth medium or culture medium. Culture media are usually purchased from cell culture supply companies and contain nutrients designed for the optimal growth of the type of cells the scientist wants to study. Different types of media are used to grow different types of cells. For growing bacteria or other microorganisms, for example, nutrient broths or agar plates are often used.

A complete medium designed for growing animal and human cells contains nutrient requirements, including sugars (usually glucose), amino acids, vitamins, and trace elements that are needed for the optimal growth of these cell types. In addition, culture medium might also contain a growth promotion factor called serum. Serum promotes healthy growth in cultured cells in several different ways. First, it binds to toxic chemicals to prevent them from harming the cultured cells. Serum also renders trypsin and other protein-digesting enzymes inactive, preventing them from digesting the cells. Finally, serum also contains hormonelike growth factors that the cells need. Scientists also believe that serum interacts with the growth surface (the microscope slide or the petri dish) in some manner, but they are not certain how this works. In addition to serum, most

cell culture medium also contains antibiotic and antifungal components. Antibiotics and antifungals are not necessary for growth of the cells, but they prevent contamination of the culture by unwanted bacteria or fungi.

Bacteria and fungal spores that can contaminate a cell culture are everywhere: on the fingertips of lab technicians, on countertops, and even in the air. The key to preventing contamination is the use of the aseptic, or sterile, technique. The sterile technique is a set of steps that, if carried out properly, minimize the risk of contamination and cell death. Some sterile technique methods include keeping culture flasks open for the minimum amount of time possible, holding flasks at a 45-degree angle to minimize the possibility of dust falling into them, and cleaning work surfaces with a solution of 70% ethanol.

EQUIPMENT USED IN CELL CULTURE

Tissue culture laboratories contain specialized equipment to ensure the optimal growth of healthy, uncontaminated cell lines. Cells must be grown in sterile, nontoxic containers. The containers must be biologically inert, meaning that they do not contain chemicals that would react with or harm the cells. They also need to be optically transparent so the cells can be seen through the culture vessel using a microscope. Laboratory supply companies make several types of culture vessels including petri dishes, multi-well plates, and screw top flasks. Most vessels are made of specially treated polystyrene plastic.

A tissue culture laboratory also contains biological safety cabinets called laminar flow hoods. These hoods are designed to protect the cells being cultured, laboratory personnel, and the environment from biohazards and cross contamination during routine procedures. Laminar flow hoods come in two types—horizontal and vertical. Both types of hood provide a continuous stream of air that has passed through a series of filters. Air is brought into the hood from the surroundings and forced through a filter that removes relatively large airborne particles such as pollen, dust, and lint. The air is then passed though another filter called a high efficiency particle (HEPA) filter. HEPA filters are designed to remove tiny airborne particles, such as bacteria and some viruses, that might otherwise contaminate the cell culture. By providing a continuous stream of air at a constant velocity, laminar flow hoods supply clean air to the work surface inside the hood.

The main difference between a horizontal laminar flow hood and a vertical one is the direction of the air flow. In a horizontal flow hood,

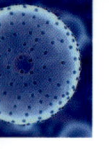

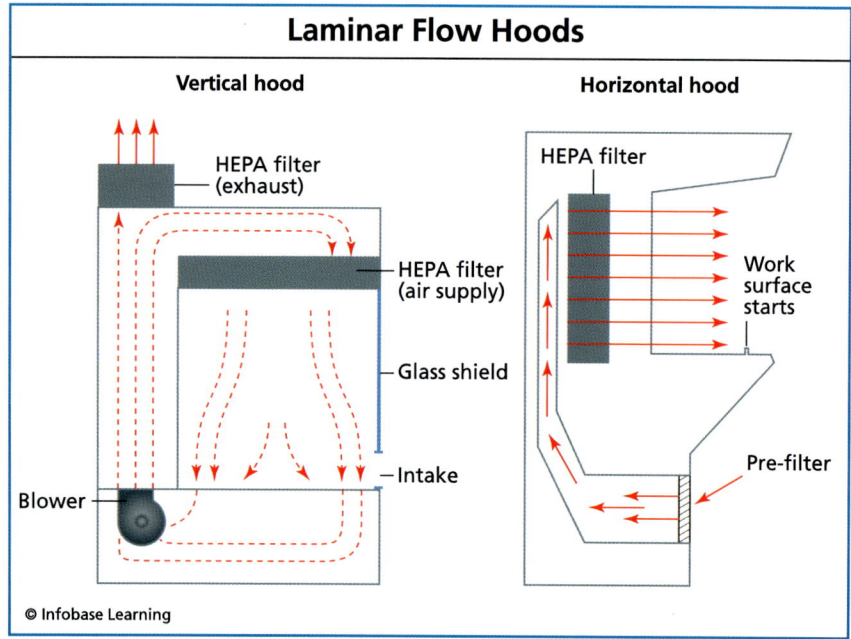

FIGURE 4.5 Laminar flow hoods are essential pieces of equipment for a tissue culture laboratory. Constantly circulating air and special filters provide a clean work surface free of bacterial and viral contamination.

filtered air is blown from the back of the hood toward the laboratory technician. This air flow keeps filtered air moving over the work surface and will help prevent contamination of the cell culture, but it is not the best choice when working with potentially hazardous organisms such as bacterial or fungal cultures. In a vertical flow hood, on the other hand, the filtered air blows down from the top of the hood. In addition, the air inside a vertical flow hood is filtered through a second HEPA filter before it is released into the room. Because of this second filter, vertical laminar air flow hoods are more suitable for working with potential biohazards.

Both types of hoods usually come equipped with an ultraviolet (UV) light source as well. The UV lamp is switched on for a few minutes before work begins in the hood. The UV light kills bacteria and fungal spores and sterilizes the hood. Technicians working with these hoods have to be careful with the UV light, however, as it can cause skin and eye damage.

In order to thrive, cell cultures must be kept at the proper temperature, moisture level, and pH. To provide these conditions, the cultures are kept in incubators with controlled temperature, humidity, and carbon dioxide

(CO_2) levels. The optimal temperature for human cell cultures is normal body temperature, 98.6° Fahrenheit (37° Celsius). To provide the correct moisture levels, most CO_2 incubators are equipped with water pans in the bottom of the incubator. If properly filled, the water pans provide the cultures with a constant humidity level. To maintain the proper pH, cells in culture require an atmosphere that is from 5% to 10% CO_2. Culture media often contains a pH indicator that will change colors when the pH is out of the optimal range. This color change provides laboratory personnel with a visual indicator that the CO_2 may need to be adjusted. To allow CO_2 into the culture flasks and other gases to escape, the lids on culture flasks are loosened. This gas exchange is crucial for the maintenance of a proper pH, which is between 7.2 and 7.5.

Whenever the doors to an incubator are opened, to either retrieve a cell sample or to place one inside the incubator, for example, CO_2, humidity, and heat escape. Automatic controls within the incubator compensate to bring the temperature (and, therefore, the humidity) and CO_2 levels back to the set points. Laboratory technicians attempt to limit the amount of time cells spend outside of their incubators to minimize the disruption of their media's pH.

Laboratory personnel use a special type of microscope, called an inverted phase contrast microscope, to view cells that are in culture. An inverted phase microscope is a type of light microscope. It increases the contrast of the transparent cells against their background by changing the way light passes through the microscope. It works because when light passes through a transparent object, the light waves slow down. This change in speed causes the light wave to shift slightly. This phase shift (in relation to the light waves that are not disrupted by a transparent object) cannot be detected by the human eye. However, it can be enhanced by a phase contrast microscope, thereby rendering the transparent cells a dark grey on a light grey background. Phase contrast microscopy was introduced by Frits Zernike (1888–1966), a Dutch physicist, in 1934. He was awarded the Nobel Prize in Physics for his invention in 1953.

HARVESTING CELLS AND SUBCULTURES

When cells are initially put into culture, they go through an inactive, or **quiescent**, phase. This growth phase is marked by little or no growth. The amount of time that cells remain in this phase depends on cell type, cell density (how many cells are in the culture vessel), growth media, and how much the cells have been handled. After the quiescent phase, cultured cells enter a phase of exponential growth. During this phase, the cells exhibit high metabolic rates, and the media they are growing in will need to be replaced so that the cells are replenished with the nutrients they need to keep growing, a process that is often called refeeding. When cells cover all the growth surfaces in a culture vessel, the cell culture is called **confluent**. Normal, nontransformed cells will stop growing when the culture vessel becomes confluent, and the number of cells will remain stationary for a time. Unless cells are removed (harvested) or the culture is split (subcultured), cells will eventually start to die off. The retardation of normal cell growth that occurs when cells become confluent is called **contact inhibition**. However, immortal cell lines do not exhibit contact inhibition. When cancerous cells cover the growth surfaces of the culture vessel, they continue to grow, piling up in mounds.

The best time to harvest or subculture cells is just before they become confluent. To subculture a cell culture means to split it up. To establish

CRYOPRESERVATION

At very low temperatures, cell metabolism slows and eventually stops. Low temperatures make possible the storing of cells for long periods of time. However, cryopreservation (freezing and storing cells) must be done carefully to prevent harming the cells. Freezing cells too fast can kill them. Harming a cell while trying to preserve it at low temperatures is called cryoinjury.

Liquid nitrogen is often used as a way to preserve cells for long periods of time. However, precautions must be taken before the cells are frozen because the formation of ice crystals inside a cell can kill it. Changes in pH, concentration of electrolytes, and dehydration can also result in cell death. To minimize the effects of freezing, cells are usually cooled very slowly. The slow cooling forces water molecules out of the cell before ice crystals can form. Most laboratories achieve the optimal rate of cooling with the use of a piece of equipment that periodically injects liquid nitrogen into the chamber containing the cells. The cells are then moved to a storage chamber that is also cooled by liquid nitrogen. The cells are generally stored at about -202° F (-130° C) because at this ultralow temperature, water crystals do not form easily.

a successful subculture, researchers need cells that have not yet become confluent and stopped growing. If cells are allowed to stop growing, they enter another quiescent phase and it may take some time before they begin to grow again. Some of these cells never start growing again. Subculturing a cell suspension simply involves removing some cells from the suspension and splitting them out into additional culture vessels. Adherent cultures, on the other hand, must be lifted from the growth surface before they can be split. This must be done without harming the cells. One method used to release cells from the growth surface is to physically scrape them off using a rubber spatula. However, this can result in significant damage to the cells. This method is more often used to harvest cells, when it does not matter whether the cell lives or dies, rather than subculture them when further growth is desired. The preferred method for subculturing cells is to use a release chemical. Common release chemicals include the proteolytic enzymes. Proteolytic enzymes are chemicals capable of breaking long chainlike proteins into smaller pieces and include the enzymes trypsin, collagenase, and pronase. This digestive action successfully lifts cells from the growth surface. However, if these enzymes are allowed to remain in contact with the cells for too long, the cell can be damaged as surface proteins are digested. The digestive action of these proteins can be stopped by the addition of complete media that contains growth serum. The serum inactivates the enzymes. Once the cells have lifted and the effects of the proteolytic enzymes have been halted, a cell counting device, called a hemacytometer, is used to count how many cells are in the sample. Cells are then diluted with the appropriate amount of fresh growth media and split into a number of culture vessels.

Cells are subcultured when additional cells are needed. Cells are collected, or harvested, when researchers want to do additional tests on them. Some types of human cells, for example, are grown in petri dishes that contain microscope coverslips. The cells grow on top of the coverslip. During harvesting, the cells are processed by adding certain chemicals in a particular order at timed intervals. The resulting fixed cells can be mounted onto a microscope slide, stained, and observed under a light microscope.

PROBLEMS IN CELL CULTURE

The media that cells grow in is nutrient rich and will allow the growth of many organisms along with the desired cell line. As discussed, contamination with bacteria, fungi, or other microorganisms can lead to

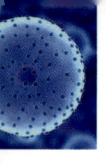
cell death. However, another problem that researchers face is the cross contamination with other cell lines. Cross contamination occurs when one or more cells from another cell line is mixed accidently with the desired cell line. Cross contamination can cause confusing test results. In a research laboratory studying the effect of a virus on a particular cell line, for example, researchers may encounter confusing results if a cell line resistant to that particular virus contaminates their cultures. In medical laboratories, cross cell line contamination could also lead to a misdiagnosis.

Contamination of a mammalian cell culture by bacteria or fungi can generally be detected with the naked eye as the liquid in the culture vessel turns cloudy. However, contamination by mycoplasma, tiny prokaryotes that lack a cell wall and have the ability to adhere to a host cell, is much harder to detect. In the laboratory, mycoplasma contamination is often discovered when previously healthy adherent cell cultures begin to lift off of the growth surface for no apparent reason. Certain types of staining methods and other biochemical processes can be performed to confirm suspected cases of mycoplasma contamination.

Contamination is not the only problem that laboratory personnel face when growing cells in culture, however. Nutrient depletion is also a problem, especially during the exponential growth period. During this period, cells are growing rapidly and using up the nutrients in their growth media at an increased rate. They also produce by-products of digestion that can prove to be toxic. Refeeding the cultures, meaning replacing the old media with new, will eliminate this problem. Refeeding also helps eliminate the accumulation of dead cells that can inhibit cell growth.

Overcoming the problems with growing cells outside of the body has provided scientists with another tool that allows them to study the structure and function of the cell and how it interacts with other cells in the body.

5

Tracking Molecules Within Cells

The membrane that surrounds a cell helps the cell keep its shape. It also protects the cell by allowing certain substances into the cell when they are needed and keeping other substances out. The membrane that allows some substances to enter a cell while keeping others out is called a **semipermeable membrane**.

OSMOSIS AND DIFFUSION

Water moves through the semipermeable membrane of a cell from regions of higher water concentration to areas of lower water concentration through a process called **osmosis**. Other small molecules (other than water) are also allowed through the semipermeable cell membrane. These molecules include oxygen, carbon dioxide, amino acids, and glucose. Like water, these molecules also move from areas of high concentration to areas of low concentration. The movement of molecules other than water though the semipermeable cell membrane is called **diffusion**. The process is the same, but only the movement of water is termed osmosis. Larger molecules, such as sucrose, starch, and proteins, are too large to get through the membrane and must be broken down into smaller molecules before they can diffuse into the cell.

In 1748, Jean Antoine Nollet (1700–1770), a French physicist, was the first scientist to demonstrate the idea that water would pass through a thin membrane from areas of high concentration to areas of low concentration.

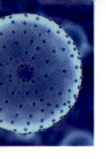

In 1824, French physiologist (René) Henri Dutrochet (1776–1847) picked up Nollet's work and, with the aid of a microscope, observed and described the movement of water into and out of plant and animal cells for the first time. Dutrochet was the scientist who named this process osmosis.

In 1854, Scottish chemist Thomas Graham (1805–1869) became the first scientist to describe how a semipermeable membrane separated substances by showing that sodium chloride and urea (the main component of urine) passed through an ox bladder. The ox bladder acted as a filter, allowing some substances to pass through while preventing others from passing. Through many experiments, Graham discovered that some substances—glue, gelatin, and starch, for example—diffused through a membrane slowly. He called these substances colloids. Other substances, such as sodium chloride, or other inorganic salts, diffused more rapidly. He called these faster diffusing substances crystalloids. Graham's studies of

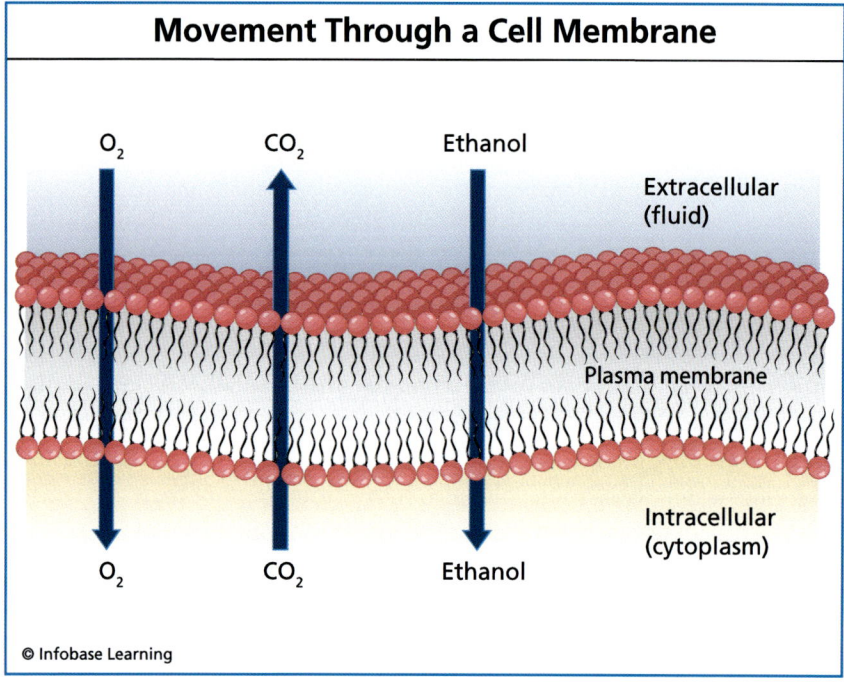

© Infobase Learning

FIGURE 5.1 The cell membrane is selectively permeable, meaning that some substances can pass through it and some cannot. Fat-soluble substances, such as oxygen, carbon dioxide, and alcohol, pass through cell membranes by simple diffusion because they can dissolve in the lipid bilayer.

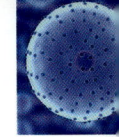

diffusion eventually lead to the development of dialysis for patients whose kidneys have failed.

In 1877, Wilhelm Pfeffer (1845–1920), a German botanist, constructed an artificial semipermeable membrane, called the Pfeffer cell, that could be used to study osmotic pressure. Using his semipermeable membrane, Pfeffer was able to precisely measure osmotic pressure for the first time. However, he was not able to develop a mathematical formula that would predict osmotic pressure. The mathematical formula that describes the principles of chemical equilibrium and osmotic pressure was worked out by Dutch chemist Jacobus Henricus van't Hoff (1852-1911) in 1885. Van't Hoff was awarded the 1901 Nobel Prize in Chemistry for this discovery. He was also the one who coined the term *semipermeable* to describe the cell membrane.

FLOW CYTOMETRY

One of the tools at the disposal of cell scientists is the flow cytometer. A flow cytometer is an instrument that can measure multiple characteristics of individual cells in a cell sample. Inside a flow cytometer is a laser beam. Every second, thousands of cells are passed through this light beam. As cells pass through the laser light, they refract, or scatter, light in all different directions. Sensors inside the flow cytometer detect the amount and direction of light scattered. The amount of forward scatter is related to cell size. Small cells produce a small amount of forward scatter whereas larger cells have a larger forward scatter pattern. Light that is scattered to the side of the cell, on the other hand, is related to the make up of the cell. The more a cell scatters light to the side, the more complex the makeup of the cell.

Another parameter that a flow cytometer can measure is fluorescence. To study fluorescence patterns with a flow cytometer, fluorescent molecules, called fluorophore-labeled antibodies, are added to the cell sample. These antibodies recognize and bind to particular molecules on the cell's surface or interior (depending on what type of **antibody** is added to the sample). When the proper wavelength laser light, called excitation light, hits the fluorophore, it emits light of a certain wavelength that is detected as a particular color of light. Different antibodies are used depending on which molecule a researcher is trying to detect. For example, one fluorescently labeled antibody might bind to DNA or RNA while a different one would be used to detect a particular protein on a cell membrane or within

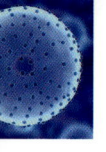

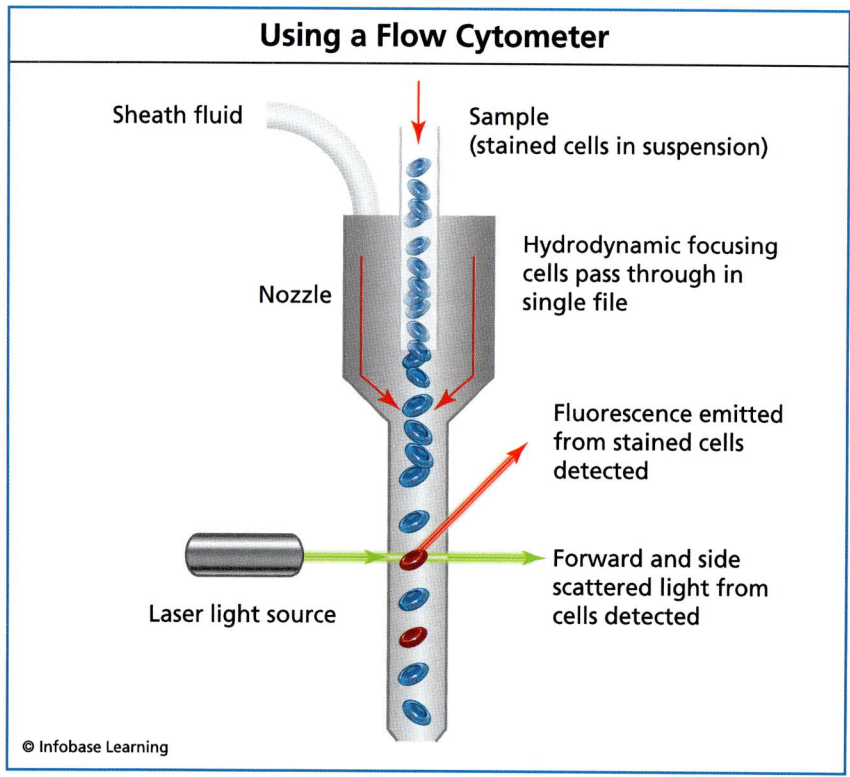

Using a Flow Cytometer

Sheath fluid

Sample
(stained cells in suspension)

Nozzle

Hydrodynamic focusing
cells pass through in
single file

Fluorescence emitted
from stained cells
detected

Laser light source

Forward and side
scattered light from
cells detected

© Infobase Learning

FIGURE 5.2 Scientists use a flow cytometer to collect data about the size, complexity, and make up of individual cells in a cell sample. Because thousands of cells can be analyzed per second, a flow cytometer greatly speeds up data collection.

the cell. The use of a flow cytometer allows scientists to collect statistical data on a large population of cells at one time.

CELL FRACTIONATION

Scientists use a variety of methods to break open cells so that they can study their structure and components. Cells can be broken down using physical methods, such as forcing the cells through tiny openings at high pressure. Chemical methods, such as adding enzymes that break down cells, may also be used. Using high frequency sound waves, a process called sonication, will also break apart cells. The process of breaking open a cell is called **homogenization**.

Each cell component—for example, the nucleus, mitochondria, and chloroplasts—all have different characteristic sizes, shapes, and densities. Scientists can use these differences to separate the different organelles from each other. The isolation of different organelles is a process called fractionation. Separating the resulting stew of unbroken cells, cell membranes, enzymes, organelles, and other cell parts requires a process called centrifugation. Centrifugation involves spinning the sample at high

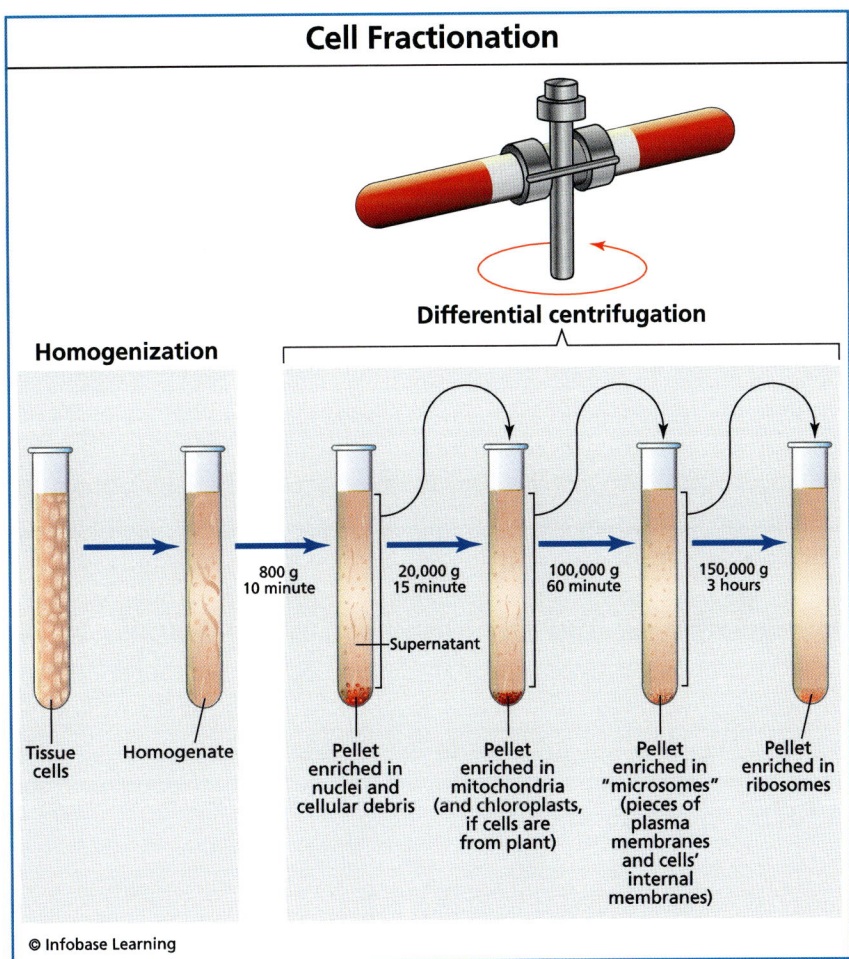

Cell Fractionation

Differential centrifugation

Homogenization

800 g
10 minute

20,000 g
15 minute

100,000 g
60 minute

150,000 g
3 hours

Supernatant

Tissue cells

Homogenate

Pellet enriched in nuclei and cellular debris

Pellet enriched in mitochondria (and chloroplasts, if cells are from plant)

Pellet enriched in "microsomes" (pieces of plasma membranes and cells' internal membranes)

Pellet enriched in ribosomes

© Infobase Learning

FIGURE 5.3 Cell fractionation is the separation of homogeneous sets, usually organelles and components, from a heterogeneous population of cells. There are two phases of cell fractionation: homogenization and centrifugation.

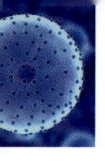

speeds. There are two methods of centrifugation that are commonly used for separating cell components—differential centrifugation and buoyant density centrifugation.

During differential centrifugation, the homogenized cell suspension is placed into a centrifuge tube. The tube is then spun at low speed for about 10 minutes. During this low-speed spin, centrifugal force drives the larger, heavier cell components, such as unbroken cells, cell walls, and flagella, to the bottom of the tube. These heavier components make up what is called the pellet. The leftover liquid above the pellet contains smaller, less dense cell components and is called the supernatant. The smaller, less dense cell components in the supernatant can be separated even further by centrifuging them again at a higher speed for a longer period of time (about an hour). During this longer, higher-speed spin, the larger, more dense components, such as ribosomes and particles of cell membrane that were left in the supernatant, are again pushed to the bottom of the centrifuge tube to create a pellet. Enzymes and other dissolved proteins remain in the supernatant.

Differential centrifugation can quickly separate some cell components from others. However, this method does not yield layers of pure components. To get a sample that contains only one type of cell component requires another method that is called buoyant density centrifugation. In this method, some of the cell suspension is added to the top of a tube that contains a sucrose (sugar) solution. The sucrose solution has a low density near the top of the tube and a high density at the bottom of the tube. This difference in density from the top of the tube to the bottom is called a density gradient. As the cell suspension is centrifuged through the density gradient, components are separated by their relative density. As the heaviest components move toward the bottom of the tube, they stop when they reach a sucrose density that is equal to their own. At the end of the process, individual cell components are separated into layers to provide scientists with pure samples of organelles that can be studied to determine their structure and function.

IMMUNOCYTOCHEMISTRY

Another technique that scientists use to study cells is immunocytochemistry. This technique is used to label certain molecules in a cell. This ability is important to research scientists as well as to a number of medical procedures, including cancer diagnosis and staging (which determines how advanced the cancer is). When this technique is used on tissues (instead of individual cells), it is called immunohistochemistry.

Immunocytochemistry involves using fluorescently labeled molecules, called antibodies, to visualize proteins and other chemicals in the cell. Depending on the nature of the experiment, the antibodies may be labeled with fluorescent dyes, radioactive elements, or other detectable chemicals. Antibodies (which are also called immunoglobulins, or IgG molecules) are proteins naturally produced by the body's immune system. These proteins seek out, identify, and attach themselves to particular chemicals, called **antigens**, on foreign cells or particles. Antigens are proteins or polysaccharides (complex carbohydrates) found on the surface of a foreign cell. When a foreign particle, whether a disease-causing agent such as a bacterium or virus, or simply a piece of pollen or dust, enters the body, cells in the body's immune system make antibodies that will detect those anti-

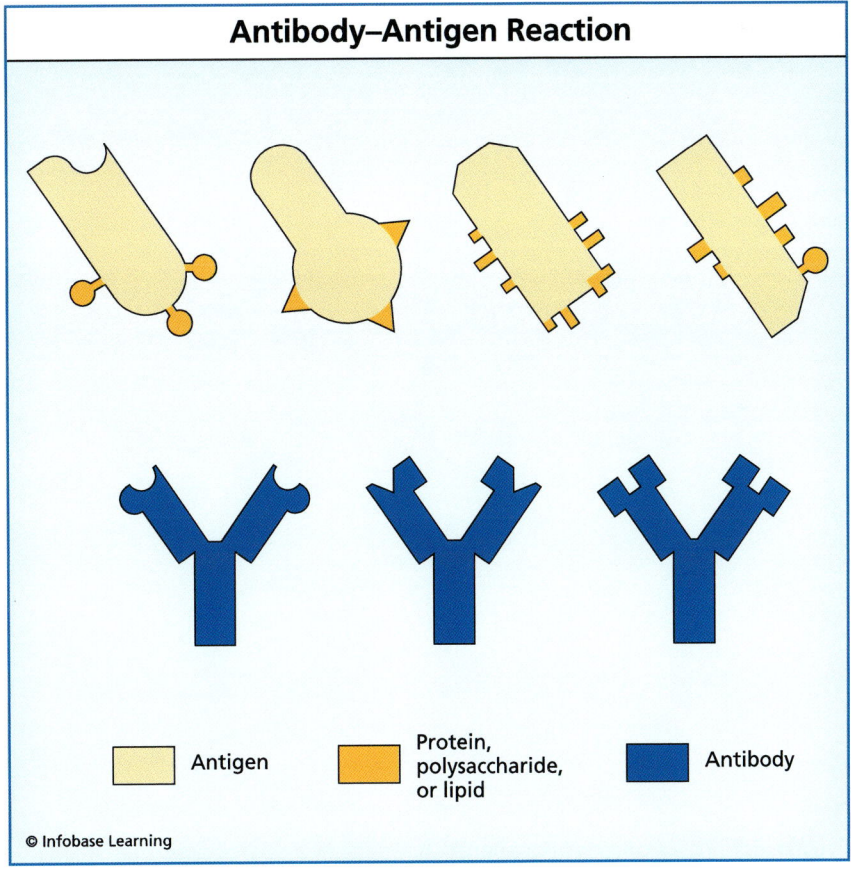

FIGURE 5.4 An antibody-antigen reaction occurs when an antibody combines with an antigen to produce an immune complex.

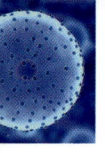

gens if it encounters it again. Antibodies can only react with the antigens they were created to detect. They will not react with other molecules.

Today, researchers can produce identical antibodies that are specific for a particular antigen in the laboratory. These antibodies, called **monoclonal antibodies**, are produced by genetically identical cultured cells (clones of a unique parent cell). Scientists use monoclonal antibodies for a number of research and medical applications.

Scientists and doctors often use combinations of specially prepared monoclonal antibodies to detect the presence of particular antigens. Each monoclonal antibody is tagged with a different fluorescently labeled enzyme. Because the different enzymes give off a different colored light

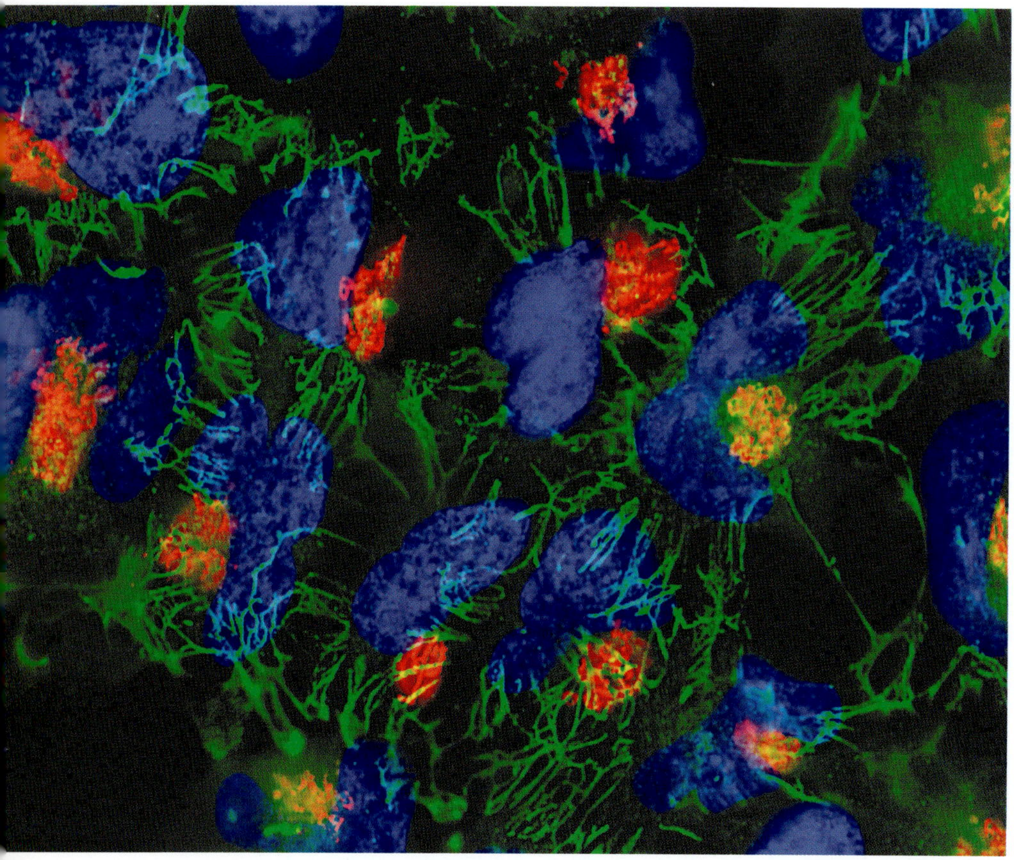

FIGURE 5.5 This immunohistochemistry image is a micrograph magnified 1,200x of HeLa cells (human cervical cancer cells), showing the cell surface receptors in green, Golgi apparatus in red, and nuclei in blue.

NANOPARTICLES AND QUANTUM DOTS

Nanoparticles are ultrasmall particles used in a variety of industries. These particles are so small that they can easily travel through the cell's semipermeable membrane. Scientists make use of this ability by attaching fluorescent dye molecules to nanoparticles. Once inside the cell, the labeled nanoparticles attach themselves to proteins. By using a laser beam that causes the nanoparticles to fluoresce, a very sensitive digital camera, and a light microscope, scientists can track the movements of the tagged proteins through the cell.

Another extremely small bit of technology that scientists use to track proteins (and the genes that code for them) makes use of microscopic semiconductor crystals, or quantum dots. Each quantum dot is a slightly different size (although they are all smaller than a millionth of an inch in diameter). When the dots are exposed to UV light, the differently sized molecules emit different colors. The largest of these crystals emits a brilliant red light while the smallest of them glows blue. In between, there are dots that emit pink, yellow, and green light, as well as many other colors. By mixing dots of different colors, scientists can create up to 40,000 labels. In the future, scientists and doctors hope to be able to use this type of technology to quickly analyze the genes of cancer patients and, hopefully, customize treatment based on their molecular profile.

when exposed to light of a particular wavelength, scientists can tell the antibodies apart.

To prepare a specimen, monoclonal antibodies are added to a cell suspension or to a tissue sample. If a particular antigen is present, the labeled antibodies lock onto them. The sample is then washed to remove excess antibodies. This step washes off all of the antibodies that did not find their corresponding antigens. By determining which antibodies have stuck to the sample, a scientist can create a detailed image of exactly which kinds of cells are present in the sample. This technique is very useful for cancer diagnosis. If a doctor removes a sample of tissue from a cancer patient's liver, for example, a pathologist would likely test the tissue with a mixture of monoclonal antibodies. Some of the

TRACKING ELECTRONS

The human body is made up of trillions of cells. Cells are made up of even smaller components called atoms. In turn, atoms are made up of even smaller building blocks called subatomic particles. There are three main subatomic particles in an atom—protons, neutrons, and electrons.

Protons and neutrons are relatively heavy particles found in the nucleus of the atom. Electrons, however, are very light and are found moving rapidly around the nucleus of the atom. It is not easy for scientists to pinpoint the location of an individual electron at any one time. Because of this uncertainty, the standard model that scientists use for an atom shows an electron cloud surrounding the nucleus. The closest scientists can get to pinpointing the exact location of a particular electron is to say that, with all probability, it was somewhere in that cloud.

In the summer of 2010, scientists in Germany and the United States successfully tracked the valence electrons in an atom for the first time. Valence electrons are the electrons on the outermost energy level of an atom (close to the outer edge of the cloud). These electrons are the ones that are involved in chemical reactions with other atoms and, therefore, give an atom its chemical properties.

Scientists have been able to track the movement of atoms inside of molecules (meaning two or more atoms that are chemically bonded together) for about 30 years. This ability has allowed researchers to see exactly how chemical bonds are made and broken by observing the process as it happens. To follow a whole atom, a laser was used to take a "snapshot" of the atom every femtosecond (10^{-15} s — a millionth of a billionth of a second). However, electrons move much faster than atoms. Therefore, to track the electrons in an atom, scientists had to invent a faster laser, one that can take snapshots about 1,000 times faster. To meet this challenge, scientists at the Max Planck Institute of Quantum Optics, located in Germany, developed an attosecond (10^{-18} s) laser. An attosecond is a billionth of a billionth of a second. One of the lead scientists in the project, Eleftherios Goulielmakis (1975–), speculates that, in the future, the information scientists learn from this laser may allow them to control chemical reactions by steering individual electrons.

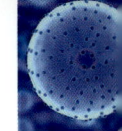

antibodies might be designed to test for liver cells, but others might specialize in detecting other types of cells, such as breast or colon cells. If the pathologist sees the antibodies designed to detect breast cells in the liver sample, it most likely means that the patient has breast cancer that has spread to the liver rather than liver cancer. This information is very important as it helps a doctor treat the patient appropriately.

Using fluorescent dyes to label antibodies was developed by American scientist Albert Hewett Coons (1912–1978) in the early 1940s. Many of fluorescent dyes he used to develop this immunofluorescent technique were developed in the same German laboratories that provided scientists with the first cell stains. Today, labeled antibodies are made in countries all over the world and come in many different colors. This development allows scientists to track several different molecules in the cell at one time.

STOCHASTIC OPTICAL RECONSTRUCTION MICROSCOPY (STORM)

Although scientists have the tools to fluorescently tag most of the proteins and genes within a cell, they have found that tagging too many proteins at one time produces a blurry image, which makes it very hard to see what is really going on. Getting a clear image of how cellular components move around the cell, interact with other cellular components, and how they work in their natural environment, could tell scientists a lot more about how cells function.

With this in mind, researchers at the University of Massachusetts Amherst are attempting to build a new microscope that will give scientists this clear image. In January 2010, physicist Jennifer Rossa and biologist Patricia Wadsworth received a grant from the National Science Foundation to help fund their research efforts. With this funding, the scientists hope that this new microscope, called the Stochastic Optical Reconstruction Microscopy (STORM), will help them identify cellular components, such as a particular protein, with greater accuracy and allow them to watch and track the proteins individually in real time. This would allow scientists to see how molecules move, interact, and how they control cellular processes. Ross and Wadsworth's technique would allow researchers to selectively turn off some of the fluorescent tags and thereby make it easier to see individual proteins. Ross and Wadsworth estimate that the new microscope will allow scientists to view molecules 100 times smaller than what is visible using current technology.

6

Genetic Engineering

Of all the molecules in the cell, the DNA molecule is a very special one. All the instructions that a cell needs to function reside within this very long molecule. The DNA molecule is shaped a bit like a spiral staircase. The two railings of the staircase are made up of sugar and phosphate. The steps on the staircase in between these two sugar-phosphate arms contain the nitrogen bases. Four types of nitrogen bases are found in DNA—adenine (A), thymine (T), guanine (G), and cytosine (C). These bases are always found in pairs. Adenine always pairs with thymine and guanine always pairs with cytosine. DNA in all living organisms has this same basic structure. Only the number of bases and their arrangement makes a human different from a dandelion.

When proteins are needed to carry out particular cellular functions, the sequence of DNA bases is transcribed into a messenger molecule called mRNA (messenger ribonucleic acid). Chemically, mRNA is similar to the DNA molecule. Its bases also consist of adenine, guanine, and cytosine. However, thymine is replaced by another base, uracil (U), in RNA. mRNA passes along the instructions coded in DNA to other cellular machinery that translates the mRNA bases into a string of amino acids. Chains of amino acids create a protein. If the base pair sequence in DNA is changed, the sequence of base pairs in mRNA is changed also. This change results in the string of amino acids being changed, too. A different order of amino acids in the growing protein molecule could result in an inactive protein or in an altogether different protein molecule.

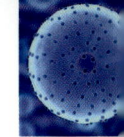

GEL ELECTROPHORESIS

One technique that scientists use to separate molecules, differently sized proteins or pieces of DNA, for example, is gel electrophoresis. The "gel" in

Structure of DNA

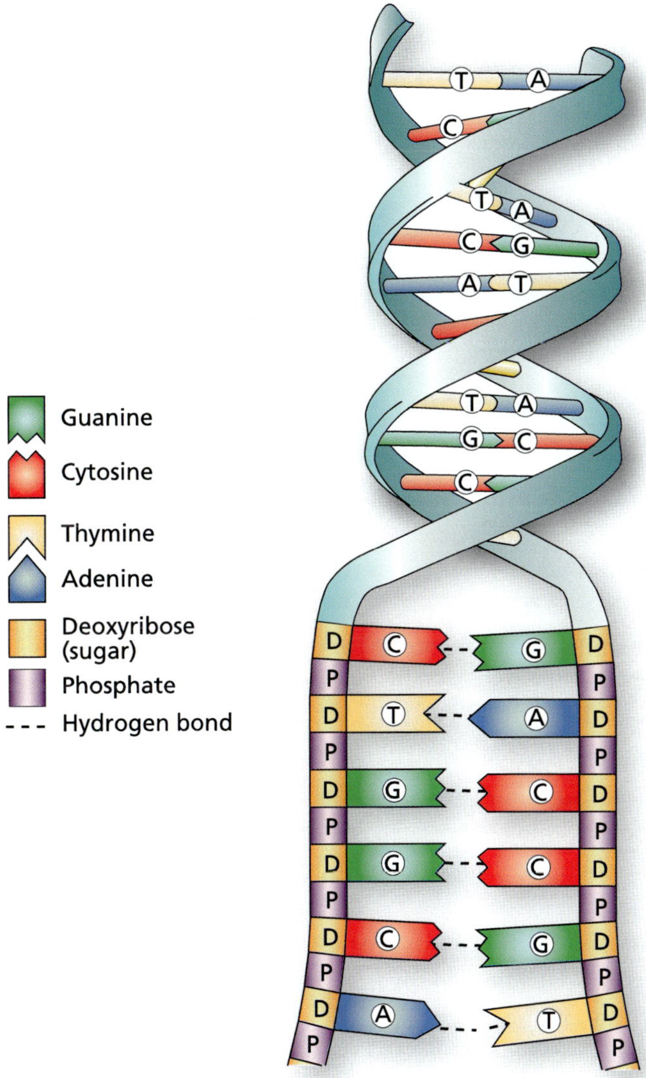

Guanine

Cytosine

Thymine

Adenine

Deoxyribose
(sugar)

Phosphate

- - - Hydrogen bond

© Infobase Learning

FIGURE 6.1 In the structure of DNA, note the base pairs of nucleotides that make up the "rungs" of the ladder.

this process is actually a type of filter with the consistency of Jell-O. The gel has many holes, like a sponge. Samples containing the molecules that need to be separated are placed at one end of the gel in small depressions, called wells. Electricity is then applied to the gel. This causes one end of the gel to become positively charged while the other end becomes negatively charged. Organic molecules, such as DNA and proteins, are also electrically charged. DNA, for example, has an overall negative charge. When an electrical current is added to the gel, the negatively charged DNA strands are attracted to the positively charged end of the gel. This attraction causes the DNA strands to move through the gel. Shorter strands of DNA (or smaller proteins) move through the gel more quickly than larger ones move. Therefore, the larger molecules do not move as far up the gel as the smaller ones do. Molecules of the same size move at the same speed and, therefore, end up in the same location on the gel. Specialized staining methods can then be used to visualize groups of molecules on the gel. The stained molecules show up as bands.

RECOMBINANT DNA

Another research technique, called recombinant DNA (rDNA), is used by scientists to create artificial DNA in the laboratory. This technique allows researchers to make multiple copies of a particular gene quickly. There are several reasons why they might want to do this. First, it allows researchers to study the gene and the protein product of that gene. The technique can also be used to make therapeutic gene products.

Recombinant DNA is DNA that originates from two or more sources (organisms). In 1972, Paul Berg (1926–) successfully produced the first strand of recombinant DNA by inserting viral DNA into the bacteria *Escherichia coli*. A year later, in 1973, Stanley Cohen (1935–) and Herbert Boyer (1936–) created the first recombinant DNA organism. At the time, Cohen, working at Stanford University, was studying plasmids. Plasmids are circular pieces of DNA found in bacteria in addition to their chromosomal DNA. Plasmids are occasionally exchanged between bacteria cells. In 1970, Cohen discovered that *E. coli* could be made to accept a plasmid from another bacteria. The newly acquired plasmid made *E. coli* bacteria resistant to the antibiotic tetracycline.

At the same time, Boyer, working at the University of California at San Francisco, was studying **restriction enzymes**. Restriction enzymes are molecules that recognize a particular sequence of DNA base pairs and cut the DNA molecule when that sequence is detected. In other words, restriction enzymes act as DNA "scissors." Together, Cohen and Boyer

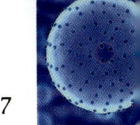

How the Immune System Works

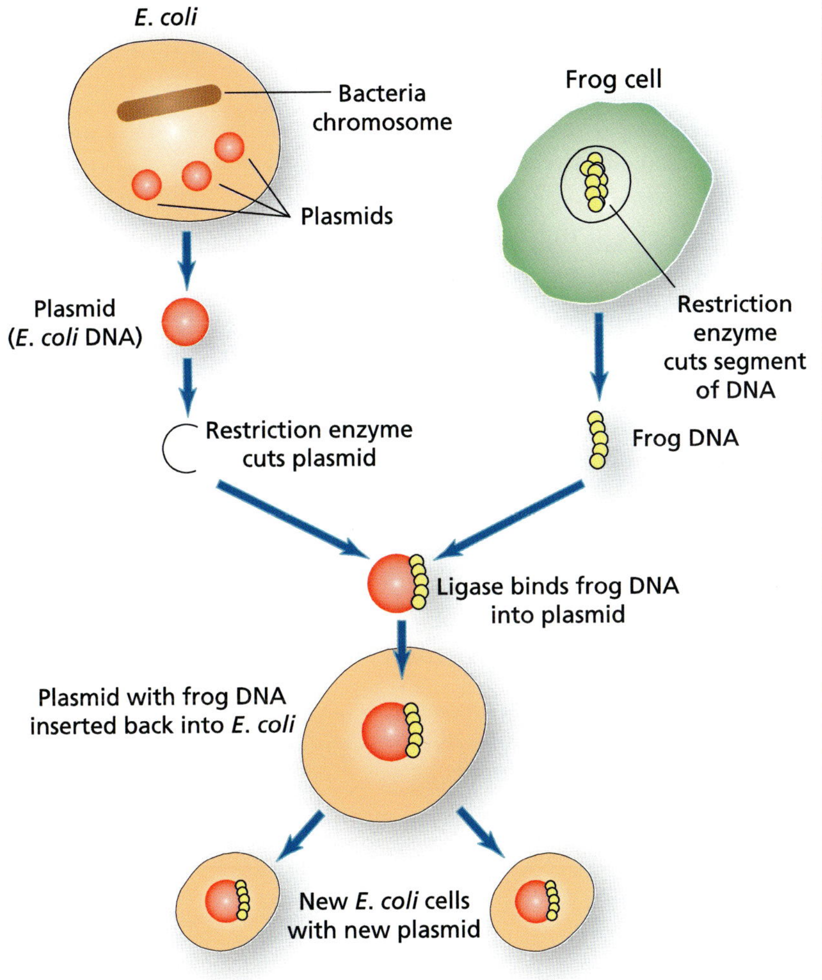

E. coli

Bacteria
chromosome

Frog cell

Plasmids

Plasmid
(E. coli DNA)

Restriction
enzyme
cuts segment
of DNA

Restriction enzyme
cuts plasmid

Frog DNA

Ligase binds frog DNA
into plasmid

Plasmid with frog DNA
inserted back into E. coli

New E. coli cells
with new plasmid

© Infobase Learning

FIGURE 6.2 In a groundbreaking gene-splicing experiment, Stanley Cohen and Herbert Boyer broke up cells of a common bacterium, E. coli, and took out small, ring-shaped pieces of DNA called plasmids. They then used a restriction enzyme to cut the plasmids open and, separately produce segments of DNA from the cells of frogs. The bacterial and frog DNA segments joined together because of the complementary "sticky ends" of single-stranded DNA attached to each segment. Boyer and Cohen used another type of enzyme, called a ligase, to bind the segments together, creating a new plasmid that contained frog and bacterial DNA. They then inserted the plasmids carrying the foreign genes into other E. coli bacteria and showed that the foreign genes could make normal proteins. When the bacteria multiplied, the added genes were duplicated along with the bacteria's own genetic material.

found a restriction enzyme that would cut the plasmid acquired by *E. coli* in a particular place. They then inserted a gene from a different bacteria into the plasmid. This foreign gene made *E. coli* resistant to another antibiotic, kanamycin. To make sure their experiment worked, Cohen and Boyer allowed the recombinant *E. coli* to multiply. Then they tested all of the bacteria's offspring for resistance to both tetracycline and kanamycin. All of the offspring survived when exposed to both of the antibiotics, making the experiment a success. Cohen and Boyer soon followed this experiment with one that introduced the genes of another species, a frog, into bacteria.

To date, scientists have identified over 100 restriction enzymes that all cut DNA in a different spot. Having all of these restriction enzymes to choose from allows scientists to select DNA sequences very precisely. During the cutting process, DNA strands with pieces of single-stranded DNA overhanging the end are created. These overhanging ends are called sticky

GENETICALLY MODIFIED ORGANISMS (GMOS)

Organisms that contain recombinant DNA are called genetically modified organisms (GMOs), genetically engineered organisms, or transgenic organisms. Many of the products that humans use today, including vaccines, some types of food, and medicines, are produced by GMOs. A majority of the genetically modified (GM) crops grown (corn, soybeans, cotton, canola, and alfalfa) now carry genes that make them resistant to particular herbicides or insects. Others include a sweet potato that is immune to a sweet potato virus, rice that contains genes that produce more iron and vitamins in the crop, and several types of crops that are now able to survive extreme weather. Fish, fruit, and nut trees that mature much faster; cows that are more resistant to mad cow disease; and pigs (called enviropigs) that excrete less phosphorus in their waste are also being created. (Excess phosphorus in animal waste can create algal blooms if it is allowed into rivers, streams, and lakes. In turn, these algal blooms create dead zones for fish and other aquatic life.)

The first GM food, a tomato designed to retain its flavor and stay fresh longer, showed up on American grocery store shelves in 1994. GM

ends. Because DNA bases (A, T, G, and C) pair up in only one way (A with T and G with C), these single-stranded pieces of DNA dictate the sequence of base pairs that can pair with it. Strands of DNA with a sequence of base pairs that matches the single-stranded are called complementary strands. When pieces of DNA with complementary ends are mixed together, they can pair with each other. Mixed in with the cut pieces of DNA is another enzyme called DNA ligase. This enzyme helps repair the DNA strands and fuse them into one piece. It acts as the "glue."

This technique can be useful in medical science. For example, scientists have been able to successfully insert the human insulin gene into *E. coli*. Bacteria multiply approximately every 20 minutes, quickly creating billions of offspring. Every time bacteria cells divide, they copy their DNA sequence. In the rDNA *E. coli*, this includes the human insulin gene recombined with its own DNA. In this way, many copies of the insulin gene are produced very quickly.

food has a variety of different advantages including less need for pesticides, fertilizers, and water as well as increased nutrition and less crop loss due to disease or insects.

However, there is some concern over the unintended consequences of growing and consuming GM food. In response to people's concerns, scientists affiliated with the Food and Agricultural Organization of the United Nations (FAO) and the World Health Organization (WHO) have conducted tests to determine the safety of GM foods. The tests have found no evidence of widespread allergic responses to the GM foods currently on the market. Another concern is gene transfer, especially of antibiotic-resistant genes, from GM foods to bacteria that could cause disease in humans. Scientists believe that the danger of gene transfer is low; however, just in case, the WHO expert panel has advised against the use of genes that confer antibiotic resistance. The movement of genes from GM foods to conventional crops has also concerned scientists. This process, called outcrossing, is a documented problem in the United States where traces of GM corn (approved only for animal feed) has appeared in corn products meant for human consumption. Agricultural leaders are looking for ways, including the clear separation of GM crops and conventional ones, to keep mixing of this kind to a minimum.

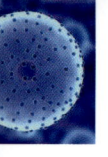

Synthetic insulin is not the only therapeutic human protein being pro-duced by transformed organisms. In fact, more than 100 rDNA products are currently on the market, including clotting factors to treat hemophilia, human growth hormone to treat some types of dwarfism, and vaccines to prevent hepatitis B infection.

DNA SEQUENCING

In 1977, Fredrick Sanger (1918–) in England and Walter Gilbert (1932–) and Gilbert's graduate student Allan Maxam (1942–), who were working independently in the United States, discovered two ways to quickly sequence the base pairs in DNA. Gilbert and Maxam's technique involved using bacteria to replicate a particular stretch of DNA. They then radioactively labeled the end of the DNA fragment to be sequenced. Restriction enzymes were used to cut the DNA strand into fragments of varying lengths. Then, gel electrophoresis was used to separate the frag-ments by length. Because the restriction enzyme cut only at a particu-lar sequence, the scientists ended up with DNA fragments with known sequences. They could then piece together the fragments to determine the sequence of the entire DNA strand they had in the beginning.

The Sanger method, on the other hand, begins with a piece of single-stranded DNA. A complementary strand of DNA is constructed with the use of an enzyme called the DNA polymerase. This sample of DNA is then further divided into four samples. Each sample is treated with one of the DNA bases (A, T, G, or C) and a chemically altered base (called dideoxy bases). The chemically altered bases are incorporated into the DNA chain like their unadulterated counterparts. However, the chemically altered bases terminate the chain. Like the Gilbert-Maxam method, Sanger's method also yielded fragments of DNA with different lengths that could be separated by gel electrophoresis. The pattern on the electrophoresis gel reveals the sequence of bases in the DNA fragment. In 1977, Sanger used this technique to sequence the genome of a bacteriophage (a virus that infects bacteria) called phi X 174. This was the first DNA genome ever sequenced. Gilbert and Sanger shared the 1980 Nobel Prize in Chemistry for their discovery of DNA sequencing. Paul Berg was also awarded part of this prize for his discovery of recombinant DNA techniques.

A year later, in 1978, David Botstein, Ronald Davis, Mark Skolnick, and Ray White proposed the idea that restriction fragment length poly-

(*continues on page 84*)

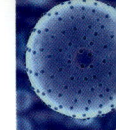

Dideoxy Chain Termination Method of DNA Sequencing

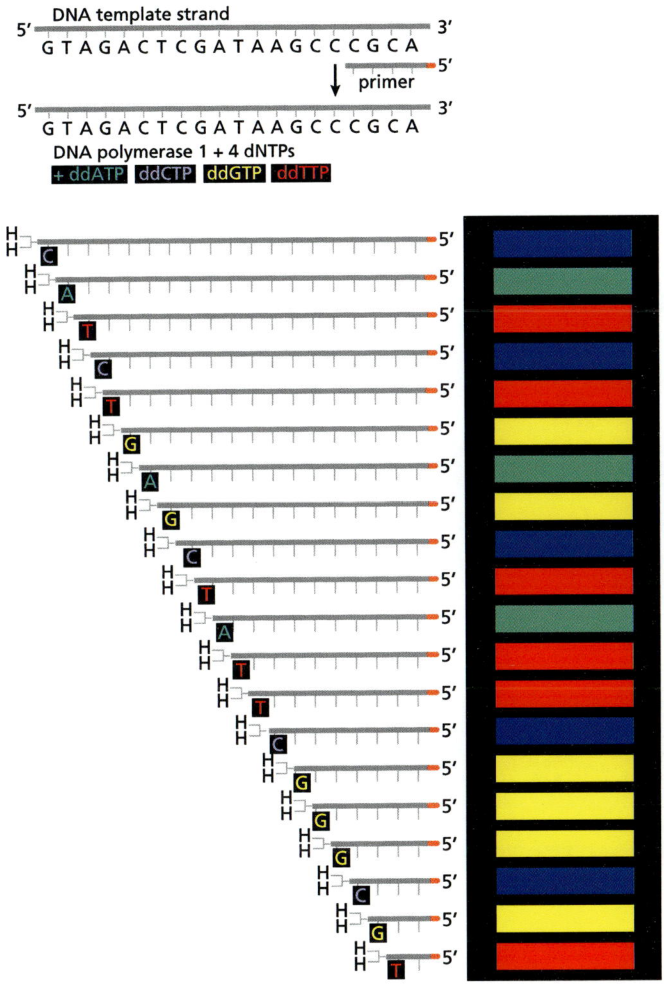

DNA template strand

5′ ———————————————————— 3′
G T A G A C T C G A T A A G C C C G C A
5′

↓ primer

5′ ———————————————————— 3′
G T A G A C T C G A T A A G C C C G C A
DNA polymerase 1 + 4 dNTPs
+ ddATP   ddCTP   ddGTP   ddTTP

© Infobase Learning

FIGURE 6.3 Copies of the DNA to be sequenced are divided into four test tubes. A molecule called a DNA polymerase begins to make a copy of the strand by picking up free bases and stringing them together. Each of the four batches contains a special version of one of the bases, called a ddNTP. At random times while copying, the DNA polymerase picks up a ddNTP instead of the normal base. This interrupts the process of copying, leaving molecules that are broken off at each position in the strand. The length of the strand and a radioactive label (or a fluorescent marker) show researchers which base is at each position.

DNA FINGERPRINTING

In 1984, British geneticist Alec Jeffreys (1950–   ) developed a method of using RFLPs to identify people. First, DNA is isolated from a sample of body tissue or fluid, which may include blood, saliva, or hair. The DNA is then cut with restriction enzymes to create RFLPs. The RFLPs are placed on an electrophoresis gel and sorted by size. Finally, a Southern blot is performed. The resulting pattern on the Southern blot looks a little bit like a bar code and this pattern is called a DNA profile or DNA fingerprint. However, using RFLPs to develop a DNA fingerprint requires a relatively large amount of DNA. Since 1984, scientists have developed other, more efficient ways of developing the same data.

Today, for example, forensics experts can use PCR to amplify even the tiniest bits of DNA left behind at a crime scene. They then probe the DNA, searching for 13 specific polymorphic sites, which are called short tandem repeats (STR), that vary from person to person to create a genetic profile. The chances that two people will have the same genetic profile for all 13 of these regions is very small (about one in a billion). As of August 2010, the national DNA databank (known as the combined DNA index system, or CODIS) contains over 8,786,734 profiles in its convicted offender index and 332,527 profiles developed from crime scene evidence. Identifying and eliminating criminal suspects is only one of the ways that DNA profiling is used. DNA fingerprinting may also be used to identify human remains, establish paternity, match organ donors with recipients, or confirm animal pedigrees.

As of August 2010, CODIS helped solve more than 100,000 criminal cases. However, storing data generated by DNA fingerprinting is not

---

Opposite: FIGURE 6.4 In the classic form of Alec Jeffreys's DNA profiling test, DNA is recovered from the nuclei of cells taken from a biological fluid, such as blood or semen. Enzymes from bacteria break the DNA into fragments. The fragments are separated according to size by a process called gel electrophoresis and placed on a membrane. The tester then adds radioactive segments of DNA, which will attach themselves to sequences in the test DNA that vary from person to person. The membrane, which contains DNA samples from the crime scene, suspect, and victim on different "tracks," is then exposed to X-ray film. When the film is developed, it reveals patterns that look like supermarket bar codes. A computer can compare the patterns to determine whether the crime scene sample came from a suspect, a victim, or someone else.

without controversy. The primary concern is individual privacy. Unlike fingerprints, which can only be used for identification purposes, DNA profiles contain much more individual information, including the predisposition to particular diseases and information relating to possible

(continues)

DNA Profiling

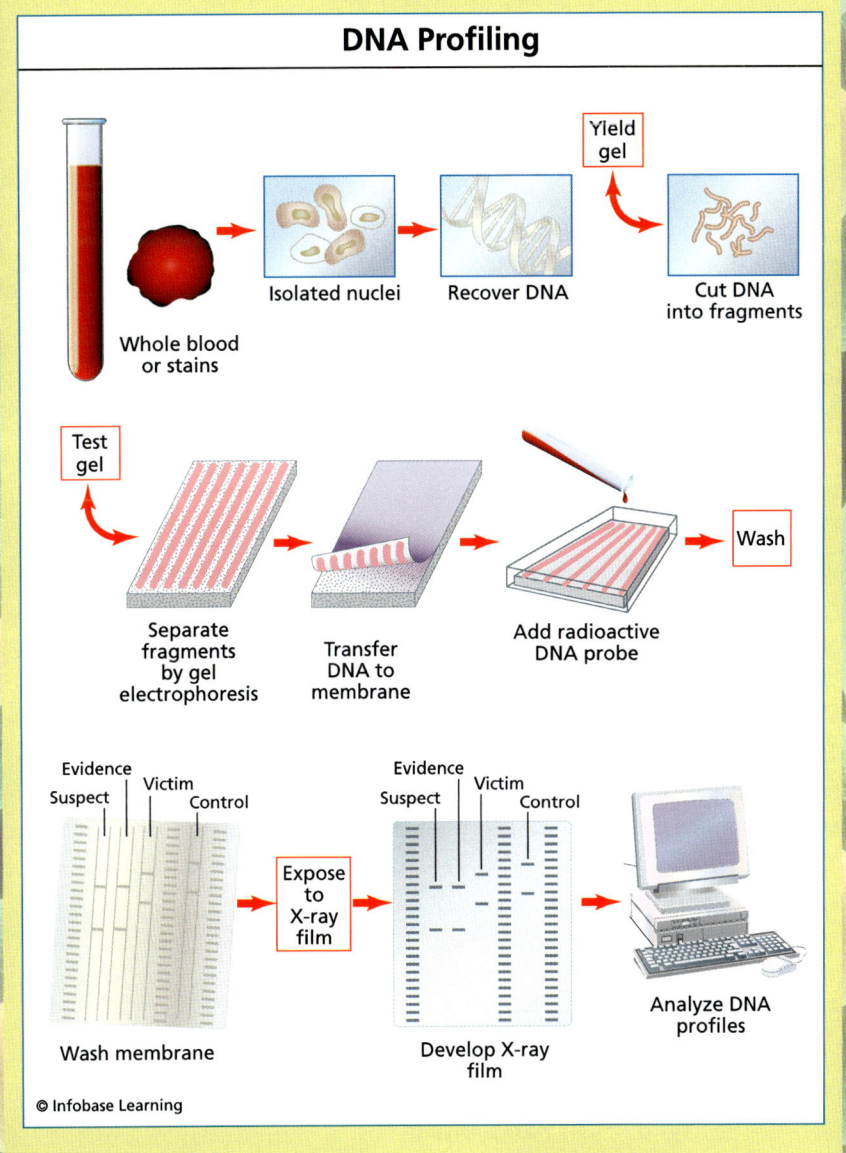

© Infobase Learning

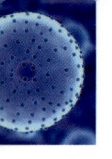

(continued)

paternity issues. If not protected properly, DNA profiles could become the basis of discrimination against individuals by insurance companies, banks, employers, and others. As of July 2008, all 50 states require convicted sexual offenders to submit to DNA testing. Forty-six states require all other convicted felons to supply DNA samples, too. In 11 states, DNA is required from people convicted of certain misdemeanors and in 12 states people who are arrested, but not convicted, in association with particular crimes (usually felonies) can be required to submit a DNA sample. Many states have no laws requiring the destruction of DNA records if arrestees are not convicted or if a conviction is overturned, possibly leading to the storage of innocent people's DNA profiles.

(continued from page 80)

morphism (RFLP) could be used in mapping genes. Nearly everyone's DNA sequence is slightly different (with the possible exception of identical twins). If a genetic variation is visible in at least 1% of the population, the variation is considered a polymorphic trait. Polymorphic traits are traits that exist in several distinct forms. For example, the ABO blood types are a polymorphic trait.

RFLPs may be inherited from a parent or they may arise due to mutation (change) in a person's genome. These differences in sequence may create or destroy the sequences recognized by various restriction enzymes. The creation or destruction of recognized restriction sites leads to DNA fragments of different lengths between different people. Because RFLPs are often inherited, they can be used to test for paternity or for specific genetic diseases. They can also be used to determine if someone is the source of a particular DNA sample.

RFLPS are detected using a method called a Southern blot. A Southern blot reveals a DNA sequence by attaching a **hybridization probe** to DNA fragments that have been separated using gel electrophoresis. A hybridization probe is a length of single-stranded DNA with a known sequence. The probe is labeled with a radioactive element or with a fluorescent dye so it can be detected. If the hybridization probe encounters its complementary DNA sequence, it will stick. Because scientists know the sequence of the

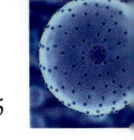

hybridization probe, they will know the sequence of the DNA strand to which it sticks. For example, if the hybridization probe has a sequence of GATTCGAAT, the DNA strand it binds to must have a sequence of CTAAGCTTA (because adenine always pairs with thymine and guanine always pairs with cytosine).

The Southern blot was developed in 1975. The technique is named for its founder, British biologist Edwin Southern (1938–   ). Scientists also use a technique known as Northern blotting (a play on Southern's name). The idea and technique behind Northern blotting is nearly identical to Southern blotting. The difference between the two is that Northern blotting is used to determine the sequence of RNA rather than DNA. Another similar technique, called Western blotting, is used to determine if a particular protein is present in a mixture of proteins. Western blotting can also reveal the protein's size.

In 1983, American biochemist Kary Mullis (1944–   ) developed a technique called polymerase chain reaction (PCR), which can duplicate fragments of DNA. The use of PCR can quickly amplify even the smallest bits of DNA, producing thousands to millions of copies of the same sequence. This mass production of identical DNA sequences makes DNA fingerprinting and sequencing much easier because scientists have more material to work with. Mullis was awarded the 1993 Nobel Prize in Chemistry for his discovery.

In 1986, Leroy Hood (1938–   ) an American biologist at the California Institute of Technology, improved on and automated Sanger's method of DNA sequencing. Sanger's method used radioactive labels, which Hood found too unstable for the automated process. The radioactivity also posed a health risk for scientists working with the probes. To decipher the DNA sequence, Sanger's method also required the time-consuming process of running four different gels (one for each of the four DNA bases). To address these problems, Hood replaced the radioactive tags with fluorescent dyes (using a different color for each of the four DNA bases). The fluorescent dyes posed no health risk to laboratory personnel and their different colors eliminated the need to run four gels. In Hood's technique, a laser excites the fluorescent dyes, causing them to give off colored light as the DNA fragments move through the gel. A digital camera is used to pick up these signals and the data is transmitted to a computer for analysis.

As time went on and gene mapping techniques improved, a flurry of firsts began to appear in scientific publications. In 1990, scientists announced a plan to map the entire human genome by 2005. In 1995, the first bacterial genome was sequenced. Three years later, in 1998, the genome of the first multicellular organism, a microscopic worm called a nematode, was sequenced. In 2000, the first plant genome was sequenced and the private company, Celera Genomics, led by Craig Venter, announced that it had finished sequencing the entire human genome. The publicly funded Human Genome Project also announced that they had constructed a "working draft" of the human genome that same year. Both versions of the human genome were published in the scientific journals *Science* and *Nature*. Independently, scientists at Celera and at the Human Genome Project determined that the human genome is made up of approximately 30,000 genes. The complete draft of the human genome was released in 2003, and the Human Genome Project was declared complete.

GENETIC ENGINEERING

With advances in DNA sequencing, scientists have learned more and more about how DNA functions in the cell. This knowledge has led to experiments that manipulate DNA in various ways. For example, scientists have successfully created organisms with genes that have purposefully been eliminated or made inoperable. Making a gene (which ultimately codes for a protein) inoperable can tell scientists a lot about the gene's function. The process of eliminating copies of a gene from an organism's genome is called a gene knockout. Because mice and humans share a number of genes, scientists use knockout mice to help them study different kinds of human diseases, including cancer, obesity, heart disease, aging, diabetes, and a number of others.

Genetic engineering can also be used to make copies of organisms. Making an exact genetic replica of an organism is called cloning. Scientists can make an exact replica of an organism in two ways—artificial twinning and somatic cell nuclear transfer. In nature, a fertilized egg grows by splitting into two cells. Those two cells split into four. Four cells grow into eight and so on. Identical twins result when the first two cells separate and begin to grow on their own. Because they originate from the same fertilized egg, identical twins have (nearly) the same DNA sequence. Artificial twinning is very similar. The difference lays in the fact that a scientist separates the two cells in a petri dish instead of allowing them to separate on their own. The separated eggs are then implanted into a surrogate mother to develop as normal.

Somatic cell nuclear transfer (SCNT), on the other hand, does not begin with a fertilized egg. Instead, the DNA to be cloned is taken from a somatic cell. Somatic cells are any body cells except for eggs and sperm. In

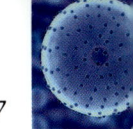

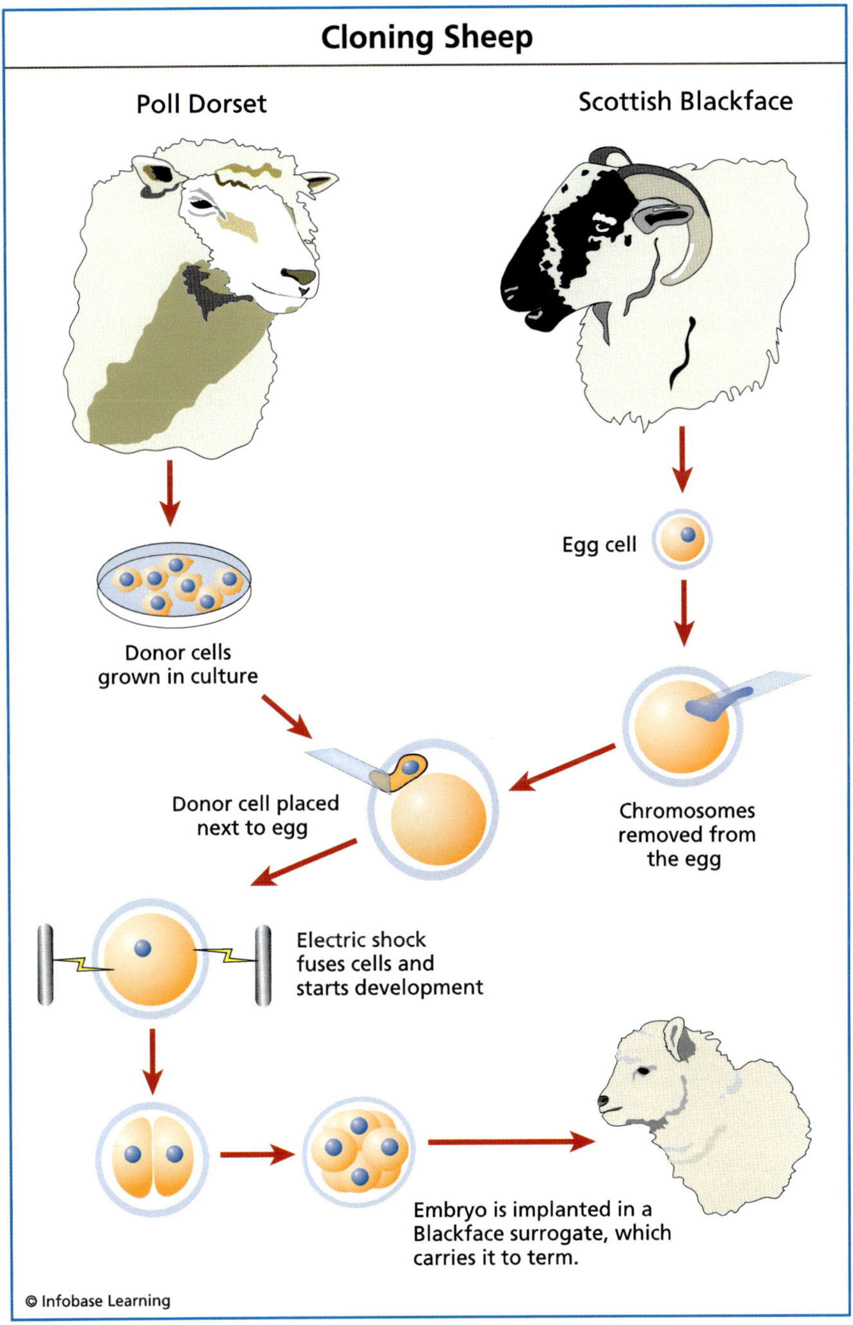

Cloning Sheep

Poll Dorset   Scottish Blackface

Egg cell

Donor cells
grown in culture

Donor cell placed
next to egg

Chromosomes
removed from
the egg

Electric shock
fuses cells and
starts development

Embryo is implanted in a
Blackface surrogate, which
carries it to term.

© Infobase Learning

FIGURE 6.5 In this schematic of how sheep can be cloned, the Poll Dorset
sheep provides the nucleus, which is obtained from cultured ovine mammary
gland epithelial (OME) cells. The Scottish Blackface provides the egg,
from which the nucleus is subsequently removed. If the cloning process is
successful, the clone will look like a Poll Dorset.

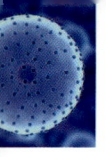

1996, Scottish scientists used SCNT to create a cloned sheep they named Dolly. To make Dolly, the scientists first removed an udder cell from a six-year-old adult female sheep. Then, they removed the cell's nucleus, including its DNA. The nucleus was transferred into an egg cell (from a different sheep) that had been enucleated (which means the cell's nucleus had been removed). By adding a variety of chemicals and electricity, the scientists successfully got the egg cell with the adult udder nucleus to begin acting like a fertilized egg. After allowing the cell to divide a few times, they then implanted the developing clone into a surrogate mother. Out of the 277 cells that were subjected to this treatment, only one cell survived to be brought to term. The result, Dolly, was an exact genetic replica of the sheep whose udder cell was taken.

Dolly was the first mammal to be cloned from an adult cell. However, Dolly may not have been completely healthy. She was only six years old when she died. (Sheep have a normal life span of 11 to 12 years.) This ignited a worldwide controversy about whether or not cloned animals die prematurely. However, there was no conclusive proof that Dolly's early death had anything to do with the fact that she was cloned. At her death, an autopsy reveled that she had died from a lung tumor caused by a virus. However, she also had arthritis in her hind legs, an affliction that is normally found in older sheep. Other than that, Dolly appeared healthy. Scientists have also cloned other animals, including frogs, mice, cattle, and monkeys.

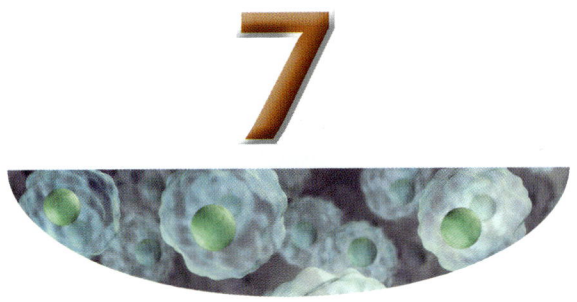

Other Research Techniques

Microscopy, tissue culture, gel electrophoresis, and recombinant DNA are just a few of the ways that scientist study cells. They also use models, biochemical assays, and many other research techniques.

MODELING

Discovering the shape of the DNA molecule required the work of many scientists and several different experimental techniques. The journey toward the final discovery began in 1868 when the Swiss physician Johannes Friedrich Miescher isolated DNA from the nuclei of white blood cells for the first time. Miescher called the substance he had isolated nuclein. However, over time, as other scientists discovered more about this macromolecule, the name was changed first to nucleic acid and then finally to deoxyribonucleic acid.

In 1919, Russian-American biochemist Phoebus Levene (1869–1940) discovered that DNA was made up of units called nucleotides. A nucleotide is made up of a phosphate group, a sugar molecule, and a nitrogen-containing base. He also discovered that the sugar in DNA was deoxyribose.

In 1944, Oswald Avery (1877–1955), a Canadian-born American physician, suggested that the DNA macromolecule carried genetic information from generation to generation. Five years later, after reading Avery's paper, Austrian-American biochemist Erwin Chargoff (1905–2002) became interested in the molecule. Expanding on Levene's work with

89

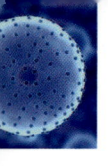

biochemical tests, Chargoff noticed that although the arrangement of base pairs in DNA varied widely between organisms, the amount of adenine in a DNA molecule almost always equaled the amount of thymine. In addition, the amount of guanine was nearly equal to the amount of cytosine. This observation is now known as Chargoff's rule. Chargoff did not himself state the base pair makeup of DNA. However, his data was one more piece of the puzzle that eventually helped James Watson (1928–) and Francis Crick (1916–2004) determine the structure of the DNA molecule.

Another piece of the puzzle came from the laboratory of American chemist Linus Pauling (1901–1994). In 1948, Pauling discovered that some proteins have alpha-helical structures. (An alpha helix looks a bit like a coiled spring.) Later, Pauling would propose that the DNA molecule had a triple-helical structure.

While Pauling's idea would turn out to be incorrect, more pieces of the structural puzzle were provided by English scientists Maurice Wilkins (1916–2004) and Rosalind Franklin (1920–1958). At the time, Wilkins and Franklin were using a technique called X-ray crystallography (also called X-ray diffraction) to try to determine the physical structure of DNA. X-ray diffraction is often used to discover how crystals are constructed. This is done by beaming X-rays through a crystal. As the X-rays encounter structural details in the crystal, some of them are scattered. This scattering pattern reveals the crystal's structure. DNA, and some other types of biological molecules, can form crystals when they are treated with certain chemicals. By treating DNA molecules and exposing them to X-rays, Franklin and Wilkins hoped to discover important clues that could tell them what the DNA molecule looked like.

In 1953, Watson and Crick saw one of Franklin's X-ray crystallography pictures of DNA. They believed that the photograph clearly showed that the DNA molecule was a helix. Using cardboard cutouts of each of the components of DNA, they tried to figure out a way to fit them all together that would explain the experimental data gathered by so many different scientists. Initially, they had a little trouble because they had an incorrect atomic structure of the thymine and guanine molecules. Then, on the suggestion of Jerry Donohue (1920–1985), an American chemist who was temporarily sharing an office with Watson and Crick, they decided to see if an alternative atomic structure of guanine and thymine would work. The two scientists quickly realized that in this new configuration, thymine paired with adenine. In addition, using data developed

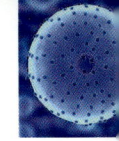

FIGURE 7.1 Discoverers of the structure of DNA James Watson (*left*) and Francis Crick (*right*) pose with their model of part of a DNA molecule in 1953.

by Pauling, they realized that the length of the adenine-thymine bond was exactly equal to the length of a guanine and cytosine bond. Because the bond lengths were the same, the base pairs would keep the sugar-phosphate chains at a constant distance from each other. They would also obey Chargoff's rule.

Once these preliminaries were worked out, Watson and Crick replaced their cardboard cutouts with a stick-and-ball model. Using this model, they worked out that the DNA molecule would fit all of the experimental data if it were a double helix made up of two sugar-phosphate chains with base pairs in the center. Watson and Crick announced their discovery in the April 1953 issue of the science journal *Nature*. The same issue also

(*continues on page 94*)

ANIMAL MODELS

The processes that go on in an individual cell do not always tell scientists how cells will work together in a whole, living body. When cells work together, things become much more complex. To advance the scientific understanding of how the body works, study disease processes, and to develop and test drugs to treat those diseases, scientists cannot experiment on human beings. Therefore, they sometimes use animals as models.

One of the more common animals used in genetics laboratories, for example, is the *Drosophila melanogaster*, a type of fruit fly. The fruit fly makes an excellent animal model because it is easily kept in the laboratory and it reproduces quickly. Studying these tiny flies helped scientists discover exactly how DNA is replicated and transcribed. Transgenic fruit flies can also help scientists understand the affect of genetic changes on development. *Caenorhabditis elegans*, a microscopic worm, is also used as an animal model. Even though *C. elegans* has a very small, simple nervous system, it carries out many of the same functions as a larger, more complex nervous system. Therefore, it has been useful for helping scientists understand how the nervous systems of more complex animals function. Genetically modified animals, like the knockout mice, can also be used as an animal model for human disease.

In 1962, Japanese marine biologist Osamu Shimomura (1928–   ), who was working at the Marine Biological Laboratory in Woods Hole, Massachusetts, isolated a protein from the *Aequorea victoria* jellyfish. When exposed to UV light, the protein glows bright green, earning it the name green fluorescent protein (GFP). Since then, GFP has been inserted into animal models to enable scientists to visualize the spread of cancer, the development of nerve cells, and to study diseases such as Huntington's disease. This is done by introducing the GFP gene into the animal's genome along with the particular gene they are studying.

Opposite: FIGURE 7.2 A transgenic green pig is pictured with a normal pig in Taipei, Taiwan. In 2006, a research team at National Taiwan University succeeded in breeding three male green pigs by injecting fluorescent green protein into embryonic pigs. There are partially green pigs elsewhere in the world, but these three pigs are the first ones that are green inside and out, including their hearts and internal organs.

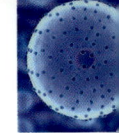

Shimomura shared the 2008 Nobel Prize in chemistry for his discovery and went on, with the help of his colleagues and other scientists, to discover several other proteins that fluoresce different colors and can be used for different procedures.

(*continued from page 91*)
included articles by Wilkins and Franklin that provided X-ray crystallography evidence of the structure.

Watson, Crick, and Wilkins were awarded the 1962 Nobel Prize in Physiology or Medicine for their work. Many people (including James Watson) have argued that Franklin should have also been awarded part of that prize. However, Franklin died of ovarian cancer in 1958 at the age of 37, and the Nobel Prize has only been awarded posthumously twice in its history (once in 1931 and again in 1961). Today, the Nobel Prize rules state that they may only be awarded to living scientists (this rule was put into place in 1974). In 2008, the records for the 1962 Nobel Prize were released and revealed that Franklin was never nominated.

COMPUTER MODELING

Stick-and-ball models and animal models are not the only tools scientist have at their disposal. In fact, one of most important and useful tools research scientists have today is the computer. Computer modeling of biological functions, called biomodeling, might, for example, help a scientist work out how the atomic structure of a particular enzyme allows the enzyme to function as it does. In turn, this understanding may help them unravel how a nonfunctioning enzyme might lead to disease. Without the use of very powerful supercomputers, the sequencing done in the Human Genome Project would not have been possible either (at least not in any kind of timely manner).

Scientists also hope that new computer models will one day help them understand how infectious diseases and cancer work at the cellular level to allow better treatments options to be developed for these diseases. To this end, Yuan-Ping Pang, a researcher at the Mayo Clinic, and his colleagues have developed computer models of proteins and enzymes related to malaria, avian flu, and severe acute respiratory syndrome (SARS). In Pang's laboratory, researchers can view a wall-sized, three-dimensional image of these molecules. Pang has already used these computer models to identify a unique amino acid sequence present in mosquitoes that carry the malaria parasite. This unique sequence could be used as a target for pesticides that would only kill these mosquitoes (and also kill the German cockroach, which also has the same amino acid sequence). Pang has also discovered an enzyme that enables the SARS virus to multiply. Current anti-SARS drug research includes hunting for drugs that can inactivate this enzyme. The supercomputer that Pang uses in his research is capable

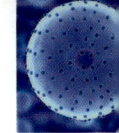

of processing one trillion calculations per second. Using this computer, with its 1,060 processors, Pang requires approximately 20 days to calculate and present an animation of the structure of a protein. If he were to use a conventional desktop computer instead, the same process would take about 28 years.

The pharmaceutical industry also uses supercomputers for a number of different processes during drug research and development. For example, computer models can help scientists determine which drugs might work. Computer-controlled robots can be used to carry out routine, repetitive testing, which frees up scientists to attend to other tasks. Robots also work more rapidly than human hands can manage. Computers also allow scientists and doctors to gather and analyze large amounts of data generated by clinical trials, an important step in the drug approval process. Collaboration between scientists has also become easier with the advent of computers. However, even with the advances in computer technology, the development of new drugs still takes an average of 12 to 15 years.

BIOCHEMICAL ASSAYS

Biochemical assays are tests run in a laboratory to look for biomarkers. A biomarker is an anatomical or biochemical feature associated with a particular biological state. Biomarkers (sometimes just called a marker) may be used to look for signs that indicate a particular disease is present or to determine the severity of disease. Tumor markers, for example, are substances given off by cancerous tumor cells. Depending on the type of tumor, these markers may be detected in blood, urine, or body tissues. The concentration of these markers can indicate various stages of disease or other conditions. The presence of the marker BTA (bladder tumor antigen), for example, could indicate that a patient has a urinary tract infection, kidney stones, or bladder cancer. A doctor would likely perform further tests to determine which condition the patient has and prescribe an appropriate treatment.

Biomarkers can also be used to detect other physiological parameters, such as the concentration of certain drugs in the body. They are used in research, too. For example, the protein Oct-4 is present only in embryonic stem cells and can be used to identify these cells.

Many biochemical assays rely on monoclonal antibodies or antigens produced (or purified) in the laboratory. This type of test is called an immunoassay. For example, to test for the presence of the West Nile virus, a sample of blood taken from a patient suspected of having West

DRUG TESTING

Scientists, doctors, and law enforcement personnel have several different tests at their disposal to help detect drugs and their metabolites (which are products of the breakdown of a drug) in the body. For example, an enzyme-linked immunosorbent assay (ELISA) may be used. ELISA makes use of monoclonal antibodies to detect and quantify proteins, antibodies, hormones, drugs, and other molecules in a cell. Another immunoassay technique used to detect legal and illegal drugs is called an enzyme multiplied immunoassay technique, or EMIT.

Another common method is thin layer chromatography (TLC). Chromatography is a technique used to separate mixtures into their components. This method is carried out on a glass or plastic slide that has been coated in a thin layer of gel (usually a silica gel). To run the test, a sample of blood, urine, or saliva (depending on which drug is being tested for) is placed, or spotted, at the bottom of the TLC slide. A standard, or control, is spotted near the bottom of the slide, too. The slide is then placed in a solvent (ethyl acetate, alcohol, or acetone, for example). As the solvent moves up the TLC plate, components in the mixture are separated. With the addition of certain chemical detectors, specific compounds can be visualized as a color change. For example, with the addition of Duquenois reagent and hydrochloric acid to a TLC slide, marijuana and its metabolites appear as a purple color. Other chromatography methods, including gas chromatography and high-pressure liquid chromatography may also be used to detect the main ingredient in marijuana, delta-9-tetrahydrocannabinol (THC), and its metabolites. THC is only present in the body for about 24 hours. However, the metabolites of THC can easily be detected for up to 60 days after the drug was last used.

Nile virus might be mixed with the West Nile antigen. If antibodies are present in the patient's blood, they will react with, or bind to, the antigen. If an antibody-antigen complex is detected, the test is positive and the person most likely has been infected with the West Nile virus. If the antibody-antigen complex does not form, the test is negative and the patient is most likely suffering from some other ailment. Doctors can also use antigens to test that a vaccination has been successful. A successful

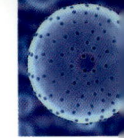

vaccination would cause the patient's body to make antibodies. If no antibodies are present, the vaccination may need to be repeated because the person is not protected against the infectious agent for which he or she is being vaccinated. Immunoassays can also be performed using antibodies to detect particular antigens. These types of immunoassays are often used to detect drugs levels, hormone levels, and cancer markers.

Because antibodies are specific to a particular antigen, they provide very promising treatments for diseases, too. For example, monoclonal antibodies that are designed to recognize a certain antigen on a cancer cell bind only to cancer cells, leaving a patient's normal cells alone. Scientists can attach a cell-killing, or cytotoxic, agent, such as a radioactive isotope, to the antibody. When the cytotoxic-antibody complex is given to the patient, the radioactive antibody searches out cancer cells in the patient's body without harming normal cells. Radiation-linked monoclonal antibodies are currently being used to kill or inhibit the growth of malignant (cancerous) cells in patients with thyroid cancer and non-Hodgkin's lymphoma.

Monoclonal antibodies can also be used to block the antigen site. For example, blocking the site that fibrinogen, a protein involved in blood clotting, normally binds to in platelets can prevent the platelets from clumping together. This is useful when a patient has had a recent angioplasty procedure (a technique to widen and unblock arteries) and clumping platelets could re-block an unblocked artery and cause a heart attack.

Biological research in general and cell research in particular are fast-moving fields of study that are relevant to everyday life. Many scientists and doctors believe that medicine will change drastically over the next several decades because of the advances made in biological research. Continued advances in computer technology will also help with future basic science research as well as with pharmaceutical research. Every year, researchers learn more about the way cells function independently and together. To make these discoveries, science tools and techniques are constantly evolving, too.

Glossary

adherent culture   A type of cell culture in which cells grow attached to a surface

antibody   Proteins produced by the immune system that bind to a specific antigen

antigens   Chemicals on the surface of a foreign cell that stimulate the immune response

cell theory   The idea that all living beings are made up of cells and that all cells originate from other cells

chloroplasts   Plant organelles in which photosynthesis takes place

clones   The genetically identical offspring that originate from one parental cell

confluent   A cell culture in which cells cover all the growth surfaces in the culture vessel

contact inhibition   The halting of cell growth that occurs when two or more cells touch

continuous cell lines   Immortal cells that can be grown indefinitely in tissue culture

deoxyribonucleic acid (DNA)   A macromolecule that is passed down from generation to generation and contains the genetic instructions a cell needs to function correctly

diffusion   The movement of molecules (other than water) through a cell's semipermeable membrane from areas of high concentration to areas of low concentration

endoplasmic reticulum   A network of membranes inside the cell cytoplasm

fractionation   A process used to separate parts of the cell

Golgi complex A complex organelle that stores, sorts, and packages proteins for use inside the cell or for export

growth medium The nutrient-rich liquid in which cultured cells are grown

homogenization A process that breaks apart cells and releases organelles

hybridization probe A length of DNA (or RNA) with a known sequence that is used to detect the complementary sequence within a DNA sample

in vitro A procedure performed in a controlled environment outside the body

in vivo A procedure performed on a living organism

microtome An instrument that slices tissue into thin pieces for microscopic examination

mitochondria Small energy-producing organelles in the cell

monoclonal antibodies Antibodies cloned from a single parent cell in a laboratory that are designed to detect a particular antigen

nucleus The part of the cell that contains DNA or RNA

objective lens Microscope lens closest to the sample that is being observed

ocular lens Microscope lens in the eyepiece

oil immersion lens An objective lens that requires a drop of special oil between it and the slide

organelles Areas inside a cell where chemical reactions necessary to cell function are carried out

osmosis The movement of water through a cell's semi-permeable membrane from regions of higher water concentration to areas of lower water concentration

quiescent Marked by inactivity or rest

refeeding The process of changing nutrient-depleted culture media with fresh media

restriction enzymes Enzymes that cut DNA at a particular nucleotide sequence

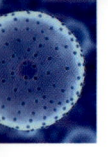

semipermeable membrane   A membrane that allows the passage of some molecules but not others

subculture   To split cultured cells into multiple culture vessels

suspension culture   Cultured cells grown while floating in culture medium

transformed cells   Normal cells that have become immortal through contact with viruses, radiation, chemicals, or oncogenes

wavelength   The distance between the crest of one wave and the crest of the next

Bibliography

American Association for Clinical Chemistry. "Laboratory Methods." Available online. URL: http://labtestsonline.org/understanding/features/methods.html. Accessed Oct. 6, 2010.

American Cancer Society. "Specific Tumor Markers." Available online. URL: http://www.cancer.org/Treatment/UnderstandingYourDiagnosis/Exams andTestDescriptions/TumorMarkers/tumor-markers-specific-markers. Accessed Oct. 6, 2010.

Barnes, Deborah. "Research in the News: Creating a Cloned Sheep." National Institutes of Health Office of Science Education. Available online. URL: http://science-education.nih.gov/home2.nsf/Educational+ResourcesTopicsGen etics/BC5086E34E4DBA0085256CCD006F01CB. Accessed Oct. 6, 2010.

BBC News. "DNA Fingerprinting 25 Years Old." Sept. 10, 2009. Available online. URL: http://news.bbc.co.uk/2/hi/uk_news/8247641.stm. Accessed Oct. 6, 2010.

Biba, Erin. "Henrietta Everlasting: 1950s Cells Still Alive, Helping Science." Wired Magazine. Jan. 25, 2010. Available online. URL: http://www.wired .com/magazine/2010/01/st_henrietta/. Accessed Oct. 6, 2010.

The Biological Stain Commission. "The History." Available online. URL: http://www.biologicalstaincommission.org/index.html?AboutUs1. Accessed Oct. 6, 2010.

Bruckner, Monica. "Basic Cellular Staining." Microbial Life. Jan. 10. 2011. Available online. URL: http://serc.carleton.edu/microbelife/research_methods/microscopy/cellstain.html. Accessed Oct. 6, 2010.

Bruckner, Monica. "Gram Staining." Microbial Life. May 29, 2008. Available online. URL: http://serc.carleton.edu/microbelife/research_methods/micro scopy/gramstain.html. Accessed Oct. 6, 2010.

Center for Biomarkers in Imaging. "FAQs." Available online. URL: http://www.biomarkers.org/NewFiles/faqs/definition.html#Anchor-What-35882. Accessed Oct. 6, 2010.

Chemical Heritage Foundation. "James Watson, Francis Crick, Maurice Wilkins, and Rosalind Franklin." Available online. URL: http://www.chemheritage

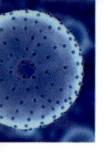

.org/discover/chemistry-in-history/themes/biomolecules/dna/watson-crick-wilkins-franklin.aspx. Accessed Oct. 6, 2010.

Chemical Heritage Foundation. "William Henry Perkin." Available online. URL: http://www.chemheritage.org/discover/chemistry-in-history/themes/molecular-synthesis-structure-and-bonding/perkin.aspx. Accessed Oct. 6, 2010.

Chudler, Eric. "Marijuana." Neuroscience for Kids. Available online. URL: http://faculty.washington.edu/chudler/mari.html. Accessed Oct. 6, 2010.

Cold Spring Harbor Laboratory's Dolan DNA Learning Center. "Biology Animation Library: Gel Electrophoresis." Available online. URL: http://www.dnalc.org/resources/animations/gelelectrophoresis.html. Accessed Oct. 6, 2010.

Cold Spring Harbor Laboratory's Dolan DNA Learning Center. "Walther Flemming." Available online. URL: http://www.dnalc.org/view/16235-Biography-7-Walther-Flemming-1843-1905-.html. Accessed Oct. 6, 2010.

Cole-Parmer. "CO_2 Incubator: A Laboratory Staple." Available online. URL: http://www.coleparmer.com/techinfo/techinfo.asp?htmlfile=CO2-incubator.htm&ID=67. Accessed Oct. 6, 2010.

Conn, H.J. "A Handbook on the Nature and Uses of the Dyes Employed in the Biological Laboratory." The Biological Stain Commission. Available online. URL: http://www.archive.org/stream/biologicalstains1953conn/biologicalstains1953conn_djvu.txt. Accessed Oct. 6, 2010.

Corning. "Cell Culture History." Available online. URL: http://www.corning.com/lifesciences/us_canada/en/about_us/cell_culture_history.aspx. Accessed Oct. 6, 2010.

Federal Bureau of Investigations. "CODIS — NDIS Statistics." Available online. URL: http://www.fbi.gov/about-us/lab/codis/ndis-statistics. Accessed Oct. 6, 2010.

Florida State University. "Joseph Jackson Lister." Available online. URL: http://microscopy.fsu.edu/optics/timeline/people/lister.html. Accessed Oct. 6, 2010.

Florida State University. "Zacharias Janssen." Available online. URL: http://micro.magnet.fsu.edu/optics/timeline/people/janssen.html. Accessed Oct. 6, 2010.

Friedrich Miescher Institute for Biomedical Research. "Friedrich Miescher: Discoverer of DNA." Available online. URL: http://www.fmi.ch/about/history/friedrichmiescher/. Accessed Oct. 6, 2010.

Gao, Dayong and J.K. Critser. "Cryopreservation of Living Cells." InformaWorld. Aug. 31, 2004. Available online. URL: http://www.informaworld.com/smpp/30982493-32759354/content~db=all~content=a713554046. Accessed Oct. 6, 2010.

Genetic Learning Center. "The Evolution of the Cell." University of Utah. Available online. URL: http://learn.genetics.utah.edu/content/begin/cells/organelles/. Accessed Oct. 6, 2010.

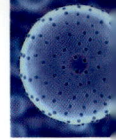

Genetic Learning Center. "What is Cloning?" University of Utah. Available online. URL: http://learn.genetics.utah.edu/content/tech/cloning/whatiscloning/. Accessed Oct. 6, 2010.

Genome News Network. "Genetics and Genomics Timeline." Available online. URL: http://www.genomenewsnetwork.org/resources/timeline/index.php. Accessed Oct. 6, 2010.

Human Genome Project Information. "DNA Forensics." June 16, 2009. Available online. URL: http://www.ornl.gov/sci/techresources/Human_Genome/elsi/forensics.shtml. Accessed Oct. 6, 2010.

Human Genome Project Information. "Genetically Modified Foods and Organisms." Nov. 5, 2008. Available online. URL: http://www.ornl.gov/sci/techresources/Human_Genome/elsi/gmfood.shtml. Accessed Oct. 6, 2010.

Invitrogen. "Introduction to Flow Cytometry." Available online. URL: http://probes.invitrogen.com/resources/education/tutorials/4Intro_Flow/player.html. Accessed Oct. 6, 2010.

Ma, Chaoyong. "Animal Models of Disease." Modern Drug Discovery. June 2004. Available online. URL: http://pubs.acs.org/subscribe/journals/mdd/v07/i06/html/604feature_ma.html. Accessed Oct. 6, 2010.

Mayo Clinic. "Visualizing the Molecules That Cause Infectious Disease: Seeing with Supercomputers." Available online. URL: http://discoverysedge.mayo.edu/de07-1-trans-pang/. Accessed Oct. 6, 2010.

Mazzarello, Paolo. "A Unifying Concept: The History of Cell Theory." Nature Cell Biology. May 1999. Available online. URL: http://www.nature.com/ncb/journal/v1/n1/full/ncb0599_E13.html. Accessed Oct. 6, 2010.

MicrobiologyBytes. "The Origins of Virology." Available online. URL: http://www.microbiologybytes.com. Accessed Oct. 6, 2010.

National Diagnostics. "Staining Procedures." Available online. Accessed Oct. 6, 2010. URL: http://nationaldiagnostics.com/article_info.php/articles_id/106

National Geographic. "Gene-Altered 'Enviropig' to Reduce Dead Zones?" March 30, 2010. Available online. URL: http://news.nationalgeographic.com/news/2010/03/100330-bacon-pigs-enviropig-dead-zones/. Accessed Oct. 6, 2010.

National Geographic. "Glowing Animals: Beasts Shining for Science." May 14, 2009. Available online. URL: http://news.nationalgeographic.com/news/2009/05/photogalleries/glowing-animal-pictures/. Accessed Oct. 6, 2010.

National Health Museum Access Excellence Resource Center. "DNA Fingerprinting in Human Health and Society." Available online. URL: http://www.accessexcellence.org/RC/AB/BA/DNA_Fingerprinting_Basics.php. Accessed Oct. 6, 2010.

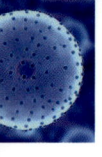
National Health Museum Access Excellence Resource Center. "Speaking the Language of Recombinant DNA." Available online. URL: http://www.access excellence.org/RC/AB/BC/Speaking_Language_rDNA.php. Accessed Oct. 6, 2010.

National Human Genome Research Institute. "Knockout Mice." Oct. 8, 2010. URL: http://www.genome.gov/12514551. Accessed Oct. 6, 2010.

National Institute of General Medical Sciences. "Inside the Cell." Available online. URL: http://publications.nigms.nih.gov/insidethecell/chapter1.html#a17. Accessed Oct. 6, 2010.

National Science Foundation. "Modeling of Biological Systems." March 14–15, 1996. Available online. URL: http://www.nsf.gov/bio/pubs/reports/mobs/mobs.htm. Accessed Oct. 6, 2010.

Nikon Microscopy U. "Introduction to Phase Contrast Microscopy." Available online. URL: http://www.microscopyu.com/articles/phasecontrast/phasemicro scopy.html. Accessed Oct. 6, 2010.

Nobelprize.org. "Glowing Proteins: A Guiding Star for Biochemistry." Oct. 8, 2008. Available online. URL: http://nobelprize.org/nobel_prizes/chemistry/laureates/2008/press.html. Accessed Oct. 6, 2010.

Nobelprize.org. "Life and Discoveries of Camillo Golgi." April 20, 1998. Available online. URL: http://nobelprize.org/nobel_prizes/medicine/articles/golgi/. Accessed Oct. 6, 2010.

Nobelprize.org. "Microscopes: Help Scientists Explore Hidden Worlds." Available online. URL: http://nobelprize.org/educational/physics/microscopes/1.html. Accessed Oct. 6, 2010.

Nobelprize.org. "Paul Ehrlich: Biography." Available online. URL: http://nobel prize.org/nobel_prizes/medicine/laureates/1908/ehrlich-bio.html. Accessed Oct. 6, 2010.

Nobelprize.org. "The Discovery of the Molecular Structure of DNA–The Double Helix." Sept. 30, 2003. Available online. URL: http://nobelprize.org/educa tional/medicine/dna_double_helix/readmore.html. Accessed Oct. 6, 2010.

Nobelprize.org. "The Nobel Prize in Chemistry 1993." Available online. URL: http://nobelprize.org/nobel_prizes/chemistry/laureates/1993/illpres/index.html. Accessed Oct. 6, 2010.

O'Connor, Clare and Ilona Miko. "Developing the Chromosome Theory." Scitable. Available online. URL: http://www.nature.com/scitable/topicpage/developing-the-chromosome-theory-164. Accessed Oct. 6, 2010.

Physorg.com. "Scientists Track Electrons in Molecules." June 13, 2010. Available online. URL: http://www.physorg.com/news195652327.html. Accessed Oct. 6, 2010.

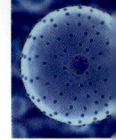

Pickrell, John. "Introduction: GM Organisms." New Scientist. Sept. 4, 2006. Available online. URL: http://www.newscientist.com/article/dn9921-instant-expert-gm-organisms.html. Accessed Oct. 6, 2010.

PopSci. "Five Reasons Henrietta Lacks is the Most Important Woman in Medical History." Feb. 5, 2010. Available online. URL: http://www.popsci.com/science/article/2010-01/five-reasons-henrietta-lacks-most-important-woman-medical-history. Accessed Oct. 6, 2010.

Purdue University. "Scanning Electron Microscope." Available online. URL: http://www.purdue.edu/rem/rs/sem.htm. Accessed Oct. 6, 2010.

Salem Press. "Osmosis." March 2006. Available online. URL: http://salempress.com/store/samples/science_and_scientists/science_and_scientists_osmosis.htm. Accessed Oct. 6, 2010.

Schiff, Judith Ann. "An Unsung Hero of Medical Research." Yale Alumni Magazine. Feb. 2002. Available online. URL: http://www.yalealumnimagazine.com/issues/02_02/old_yale.html. Accessed Oct. 6, 2010.

Science Daily. "Scientists Shed Light on Molecules in Living Cells." Aug. 27, 2007. Available online. URL: http://www.sciencedaily.com/releases/2007/08/070821081423.htm. Accessed Oct. 6, 2010.

Science Museum. "Dolly the Sheep, 1996–2003." Available online. URL: http://www.sciencemuseum.org.uk/antenna/dolly/index.asp. Accessed Oct. 6, 2010.

Scientific American. "Mereschkowsky's Tree of Life." Nov. 19, 2001. Available online. URL: http://www.scientificamerican.com/article.cfm?id=mereschkowskys-tree-of-li. Accessed Oct. 6, 2010.

Smithsonian National Museum of American History. "Whatever Happened to Polio? The Virus and the Vaccine." Available online. URL: http://americanhistory.si.edu/polio/virusvaccine/index.htm. Accessed Oct. 6, 2010.

Sumanas, Inc. "Cell Fractionation." Available online. URL: http://www.sumanasinc.com/webcontent/animations/content/cellfractionation.html. Accessed Oct. 6, 2010.

Sumanas, Inc. "Immunohistochemistry." Available online. URL: http://www.sumanasinc.com/webcontent/animations/content/immunohistochemistry.html. Accessed Oct. 6, 2010.

Swabey, Pete. "Medicinal Purposes." Information Age. Sept. 13, 2010. Available online. URL: http://www.information-age.com/channels/information-management/features/1281993/medicinal-purposes.thtml. Accessed Oct. 6, 2010.

University of California Evolution Pages. "Antony van Leeuwenhoek." Available online. URL: http://www.ucmp.berkeley.edu/history/leeuwenhoek.html. Accessed Oct. 6, 2010.

University of California Evolution Pages. "Robert Hooke." Available online. URL: http://www.ucmp.berkeley.edu/history/hooke.html. Accessed Oct. 6, 2010.

University of California Museum of Paleontology's Understanding Evolution. "Endosymbiosis: Lynn Margulis." Available online. URL: http://evolution.berkeley.edu/evolibrary/article/_0_0/history_24. Accessed Oct. 6, 2010.

University of Massachusetts Amherst. "UMass Amherst Physicist, Biologist Develop New Microscope So Powerful It Sees Individual Molecules." Jan. 6, 2010. Available online. URL: http://www.umass.edu/newsoffice/newsreleases/articles/97279.php. Accessed Oct. 6, 2010.

Wells, William. "There's DNA in Those Organelles." Journal of Cell Biology. March 15, 2005. Available online. URL: http://www.ncbi.nlm.nih.gov/pmc/articles/PMC2254739/. Accessed Oct. 6, 2010.

Wolf, Julie. "Tissue Culture Methods." University of Maryland, Baltimore County. March 2, 2010. Available online. URL: http://userpages.umbc.edu/~jwolf/method5.htm. Accessed Oct. 6, 2010.

World Health Organization. "Cell Culture Problems: Identification and Elimination." Available online. URL: http://www.who.int/vaccines/en/poliolab/webhelp/Chapter_04/4_4_Cell_culture_problems_identification_and_elimination.htm. Accessed Oct. 6, 2010.

World Health Organization. "Food Safety." Available online. URL: http://www.who.int/foodsafety/publications/biotech/20questions/en/. Accessed Oct. 6, 2010.

Further Resources

Friedman, Lauri and Marcovitz, Hal. *Is Stem Cell Research Necessary?* San Diego, Calif.: ReferencePoint Press, 2009.

Hall, Linley Erin. *Careers in Biotechnology.* New York: Rosen Publishing Group, 2007.

Harding, Lauri. *DNA Databases.* Detroit, Mich.: Greenhaven Press, 2007.

Morgan, Sally. *Super Foods: Genetic Modification of Food.* Chicago, Ill.: Heinemann Raintree, 2009.

Newton, David. *DNA Evidence and Forensic Science.* New York: Facts on File, Inc., 2008.

Pomere, Jonas. *Frequently Asked Questions About Drug Testing.* New York: Rosen Publishing Group, 2007.

Prokos, Anna. *Guilty by a Hair: Real-life DNA Matches!* New York: Children's Press, 2007.

Roleff, Tamara. *Cloning.* San Diego, Calif.: ReferencePoint Press, 2008.

Skloot, Rebecca. *The Immortal Life of Henrietta Lacks.* New York: Crown Publishers, 2010.

Web Sites

Henrietta Lacks: A Donor's Immortal Legacy
http://www.npr.org/templates/story/story.php?storyId=123232331

Learn more about Henrietta Lacks and her contribution to medical science in this NPR Fresh Air podcast with Rebecca Skloot, the author of *The Immortal Life of Henrietta Lacks.*

NOVA's "Killer's Trail"
http://www.pbs.org/wgbh/nova/sheppard/

Assemble a DNA fingerprint, identify the criminal in a hypothetical crime, and find a list of convicted inmates who have been exonerated by DNA evidence at this companion site to the NOVA program "Killer's Trail."

NOVA/Frontline Special Report: Harvest of Fear
http://www.pbs.org/wgbh/harvest/

Explore the debate over genetically modified food crops, grow your own modified and conventional crops, and take a look at the GM crops under development at this PBS Web site.

Teachers' Domain "DNA Fingerprinting"

http://www.teachersdomain.org/resource/tdc02.sci.life.gen.lp_dnamysteries/

Teachers' Domain is a resource for teachers and students. The "DNA Fingerprinting" section of the site gathers multimedia resources to help students understand the science behind DNA fingerprinting and how it is used in a criminal investigation.

Picture Credits

Index

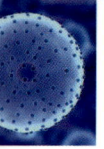

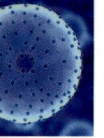

About the Author

Kristi Lew is the author of more than 30 science books for teachers and young people. Fascinated with science from a young age, she studied biochemistry and genetics at North Carolina State University. Before she started writing full time, she worked in genetic laboratories for more than 10 years and taught high-school science. When she's not writing, she enjoys sailing with her husband aboard their small sailboat, *Proton*. She lives, writes, and sails in St. Petersburg, Florida.